All Sides to an Oval

Angelo Alessandro Mazzotti

All Sides to an Oval

Properties, Parameters, and Borromini's
Mysterious Construction

 Springer

Angelo Alessandro Mazzotti
Istituto di Istruzione Superiore
"I.T.C. Di Vittorio – I.T.I. Lattanzio"
Roma, Italy

ISBN 978-3-319-81879-5   ISBN 978-3-319-39375-9   (eBook)
DOI 10.1007/978-3-319-39375-9

Printed on acid-free paper

This Springer imprint is published by Springer Nature
The registered company is Springer International Publishing AG
The registered company address is: Gewerbestrasse 11, 6330 Cham, Switzerland

To Sophie, Lilian and Anton, my greatest joy

Acknowledgements

I should like to thank Lydia Colin for her proofreading, Edoardo Dotto for his enthusiasm, Kim Williams for her encouragement and Sophie Püschmann for her technical and spiritual support.

Contents

Introduction

When writing about ovals the first thing to do is to make sure that the reader knows what you are talking about. The word oval has, both in common and in technical language, an ambiguous meaning. It may be any shape resembling a circle stretched from two opposite sides, sometimes even more to one side than to the other. When it comes to mathematics you have to be precise if you don't want to talk about ellipses, or about non-convex shapes, or about forms with a single symmetry axis. Polycentric ovals are convex, with two symmetry axes, and are made of arcs of circle connected in a way that allows for a common tangent at every connection point. This form doesn't have an elegant equation as do the ellipse, Cassini's Oval, or Cartesian Ovals. But it has been used probably more than any other *similar* shape to build arches, bridges, amphitheatres, churches and windows whenever the circle was considered not convenient or simply uninteresting. The ellipse is nature, it is how the planets move, while the oval is human, it is imperfect. It has often been an artist's attempt to approximate the ellipse, to come close to perfection. But the oval allows for freedom, because choices of properties and shapes to inscribe or circumscribe can be made by the creator. The fusion between the predictiveness of the circle and the arbitrariness of how and when this changes into another circle is described in the biography of the violin-maker Martin Schleske: "Ovals describe neither a mathematical function (as the ellipse does) nor an arbitrary shape. [...] Two elements mesh here in a fantastic dialectic: *familiarity and surprise*. They form a harmonic contrast. [...] In this shape the one cannot exist without the other." (our translation from the German, [15], pp. 47–48).

Polycentric ovals are and have been used by architects, painters, craftsmen, engineers, graphic designers and many other artists and specialists, but the knowledge needed to draw the desired shape has been—mostly in the past—either spread by word of mouth or found out by methods of trial and error. The question of what was known about ovals in ancient times has regained interest in the last 20 years. In [4] the idea of a missing chapter about amphitheatres in Vitruvius treatise *De Architectura Libri Decem* is put forward by Duvernoy and Rosin, while in [7] López Mozo argues that at the time the *Escorial* was built the know-how had to be

A.A. Mazzotti, *All Sides to an Oval*, DOI 10.1007/978-3-319-39375-9_1

better than what appears looking at the sources available today. In [6] the authors show how the oval shape has been used in the design of Spanish military defense, while in [8] the author suggests that methods of drawing ovals with any given proportion were within reach, if not known, at the time Francesco Borromini planned the dome for *S.Carlo alle Quattro Fontane*, since they only required a basic knowledge of Euclidean geometry (the whole of Chap. 7 of this book is dedicated to this construction). In any case, traces of polycentric curves date back thousands of years (see for example Huerta's extensive work on oval domes [5]). For everything that has come down to us in terms of treatises on the generic oval shape (*la forma ovata* as she calls it), with constructions explained by the architects in their own words, the book by Zerlenga [17] is a must.

The idea of this book is not to dispute whether and when polycentric ovals were used in the past. It is to create a compact and structured set of data, both geometric and algebraic, covering the single topic of polycentric ovals in all its mathematical aspects, and then to illustrate two very important case studies. Basic constructions and equations have been used and/or derived by those who needed them, but such a collection has—to the best of our knowledge—never been put together. And this is why this book can help those using the oval shape to make objects and to design buildings, as well as those using it as a means of their artistic creation, to master and optimise the shape to fit their technical and/or artistic requirements. The author went deep into the subject and the book contains many of his contributions, some of which apparently not investigated before.

The style chosen is that of mathematical rigour combined with easy-to-follow passages, and this could only have been done because basic Euclidean geometry, analytic geometry, trigonometry and calculus have been used. Non-mathematicians who can benefit from the constructions and/or formulas on display will thus be able, with a bit of work, to understand where these constructions and formulas come from.

The main published contributions to this work have been: the author's paper on the construction of ovals [8], Rosin's papers on famous oval constructions [11, 13], and on the comparison of an ellipse with an oval [11–14], Dotto's survey on oval constructions [2] and his book on *Harmonic Ovals* [3], López Mozo's paper on the various general purpose constructions of polycentric ovals in history [7], Ragazzo's works on ovals and polycentric curves [9, 10]—which triggered the author's interest in both the subjects—and finally the monograph on the Colosseum [1].

Totally new topics displayed in this monograph are *Constructions 10* and *11*, the organised collection of formulas in Chap. 4, the inscription and circumscription of rectangles and rhombi with ovals, the *Frame Problems* and the constructions of ovals minimising the ratio and the difference of the radii for any choice of axes. An organized approach to the problem of nested ovals is also presented, what the author calls the *Stadium Problem*, as well as a new oval form by the author. The main contribution is though the study on Borromini's dome in the church of San Carlo alle Quattro Fontane in Rome: together with Margherita Caputo a whole new hypothesis on the steps which Borromini took in his mysterious project for his complicated oval dome is put forward.

Chapter 2 is dedicated to the basic properties of four-centre ovals and to their proofs.

Chapter 3 lists a whole set of ruler/compass constructions systematically ordered. They are divided into: constructions where the axis lines are given, constructions where the oval has an undefined position on the plane but some parameters are given, special constructions involving inscribed or circumscribed rectangles and rhombi—including the direct and inverse *Frame Problem*—and finally an illustration of possible solutions to the *Stadium Problem*.

In Chap. 4 we prove the formulas corresponding to the constructions of ovals with given axis lines presented in Chap. 3. Then we derive the formulas solving the direct and inverse *Frame Problems*. Finally we derive formulas for the area and the perimeter of an oval.

Chapter 5 features two new optimisation problems regarding the difference and the ratio between the two radii of an oval with fixed axes and the corresponding constructions.

Chapter 6 presents famous oval shapes and their characteristics as deduced from the formulas of Chap. 4, with a final comparison illustration.

Chapter 7—co-written with Margherita Caputo—is about reconstructing the project for the dome of San Carlo alle Quattro Fontane as Borromini developed it. We suggest that he modified an ideal oval dome using a deformation module to adapt it to the space he had and to his idea on what the visitor should perceive, in terms of patterns, light and depth.

Chapter 8 illustrates a possible use of 8-centre ovals in the project for the Colosseum—as suggested by Trevisan in [16]—after a short introduction on the properties of ovals with more than four centres.

The use of freeware Geogebra by which all the figures in this book were produced, has proved itself crucial in the representation, in the understanding and in the discovery of possible properties.

References

1. AA.VV: Il Colosseo. Studi e ricerche (Disegnare idee immagini X(18-19)). Gangemi, Roma (1999)
2. Dotto, E.: Note sulle costruzioni degli ovali a quattro centri. Vecchie e nuove costruzioni dell'ovale. Disegnare Idee Immagini. **XII**(23), 7–14 (2001)
3. Dotto, E.: Il Disegno Degli Ovali Armonici. Le Nove Muse, Catania (2002)
4. Duvernoy, S., Rosin, P.L.: The compass, the ruler and the computer. In: Duvernoy, S., Pedemonte, O. (eds.) Nexus VI—Architecture and Mathematics, pp. 21–34. Kim Williams Books, Torino (2006)
5. Huerta, S.: Oval domes, geometry and mechanics. Nexus Netw. J. **9**(2), 211–248 (2007)
6. Lluis i Ginovart, J., Toldrà Domingo, J.M., Fortuny Angucra, G., Costa Jover, A., de Sola Morales Serra, P.: The ellipse and the oval in the design of Spanish military defense in the eighteenth century. Nexus Netw. J. **16**(3), 587–612 (2014)
7. López Mozo, A.: Oval for any given proportion in architecture: a layout possibly known in the sixteenth century. Nexus Netw. J. **13**(3), 569–597 (2011)

8. Mazzotti, A.A.: What Borromini might have known about ovals. Ruler and compass constructions. Nexus Netw. J. **16**(2), 389–415 (2014)
9. Ragazzo, F.: Geometria delle figure ovoidali. Disegnare idee immagini. **VI**(11), 17–24 (1995)
10. Ragazzo, F.: Curve Policentriche. Sistemi di raccordo tra archi e rette. Prospettive, Roma (2011)
11. Rosin, P.L.: A survey and comparison of traditional piecewise circular approximations to the ellipse. Comput. Aided Geom. Des. **16**(4), 269–286 (1999)
12. Rosin, P.L.: A family of constructions of approximate ellipses. Int. J. Shape Model. **8**(2), 193–199 (1999)
13. Rosin, P.L.: On Serlio's constructions of ovals. Math. Intell. **23**(1), 58–69 (2001)
14. Rosin, P.L., Pitteway, M.L.V.: the ellipse and the five-centred arch. Math. Gaz. **85**(502), 13–24 (2001)
15. Schleske, M.: Der Klang: Vom unerhörten Sinn des Lebens. Kösel, München (2010)
16. Trevisan, C.: Sullo schema geometrico costruttivo degli anfiteatri romani: gli esempi del Colosseo e dell'arena di Verona. Disegnare Idee Immagini. **X**(18-19), 117–132 (2000)
17. Zerlenga, O.: La "forma ovata" in architettura. Rappresentazione geometrica. Cuen, Napoli (1997)

Properties of a Polycentric Oval

In this chapter we sum up the well-known properties of an oval and add new ones, in order to have the tools for the various constructions illustrated in Chap. 3 and for the formulas linking the different parameters, derived in Chap. 4. All properties are derived by means of mathematical proofs based on elementary geometry and illustrated with drawings.

We start with the definition of what we mean in this book by *oval*:

> *A polycentric oval is a closed convex curve with two orthogonal symmetry axes (or simply axes) made of arcs of circle subsequently smoothly connected, i.e. sharing a common tangent.*

The construction of a polycentric oval is straightforward. On one of two orthogonal lines meeting at a point O, choose point A and point C_1 between O and A (Fig. 2.1). Draw an arc of circle through A with centre C_1, anticlockwise, up to a point H_1 such that the line C_1H_1 forms an acute angle with OA. Choose then a point C_2 on line H_1C_1, between C_1 and the vertical axis, and draw a new arc with centre C_2 and radius C_2H_1, up to a point H_2 such that the line C_2H_2 forms an acute angle with OA. Proceed in the same way eventually choosing a centre on the vertical axis—say C_4. The next arc will have as endpoint, say H_4, the symmetric to the previous endpoint w. r. t. the vertical line, say H_3, and centre C_4. From now on symmetric arches w.r.t. the two axes can be easily drawn using symmetric centres. The result is a 12-centre oval. This way of proceeding allows for common tangents at the connecting points H_n. The number of centres involved is always a multiple of four, which means that the simplest ovals have four centres, and to these most of this book is dedicated, considering that formulas and constructions of the latter can already be quite complicated. It is also true that eight-centre ovals have been used by architects in some cases in order to reproduce a form as close as possible to that of an ellipse. Chapter 8 is devoted to those shapes and to the possibility of extending properties of four-centre ovals to them.

© Springer International Publishing AG 2017
A.A. Mazzotti, *All Sides to an Oval*, DOI 10.1007/978-3-319-39375-9_2

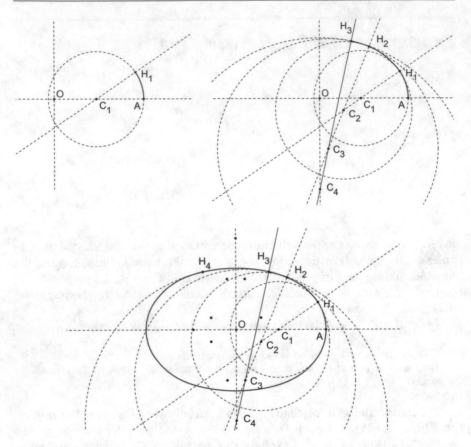

Fig. 2.1 Constructing a 12-centre oval

The above definition implies that this is a polycentric curve, in the sense that it is made of arcs of circle subsequently connected at a point where they share a common tangent (and this may include smooth or non-smooth connections). In [2] a way of forming polycentric curves using a more general version of a property of ovals is presented.

One of the basic tools by which ovals can be constructed and studied is the Connection Locus. This is a set of points where the connection points for two arcs can be found. It was conjectured by Felice Ragazzo (see [4, 5]) for ovals, egg-shapes and generic polycentric curves. A Euclidean proof of its existence was the main topic in [2], along with constructions of polycentric curves making use of it. In this book the version for four-centre ovals is displayed, along with all the implications, the main one being that the oval is a curve defined by at least three independent parameters, even when a four-centre oval is chosen.

To make this clearer let us consider an ellipse with half-axes a and b. These two positive numbers are what is needed to describe the ellipse, whose equation in a

Fig. 2.2 In any given rectangle there is only one inscribed ellipse (*the dotted curve*), and an infinite number of four-centre ovals

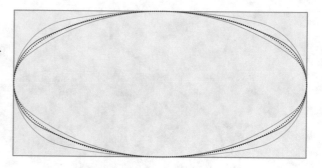

Cartesian plane with axes coincident to those of the curve and the origin as symmetry centre, is

$$\frac{x^2}{a^2} + \frac{y^2}{b^2} = 1;$$

this curve is perfectly inscribed in a rectangle with sides $2a$ and $2b$. However, a four-centre oval inscribed in the same rectangle is not uniquely determined, an independent extra parameter is needed, and in the case of eight-centre ovals the extra parameters needed are three, giving a total of five. This means more freedom in the construction and at the same time more uncertainty in the recognition of the form (see Fig. 2.2).

2.1 Four-Centre Ovals[1]

Let us devote our attention to four-centre ovals, from now on called *simple ovals* or just *ovals* (Fig. 2.3).

The double symmetry implies that it is enough to deal with the top right-hand section, a quarter-oval. In this respect, using Fig. 2.4 as reference, we define

- O as the intersection of the two symmetry axes
- A and B as the intersection points between the quarter-oval and the horizontal and vertical axes. Let $\overline{OA} > \overline{OB}$
- K and J as the centres respectively of the small and big circles, with radii r_1 and r_2, whose arcs form the quarter-oval
- H as the connecting point of the two arcs.

The above definitions and well-known Euclidean geometry theorems imply that:

---

[1]Parts of this section have already been published in the Nexus Network Journal (see [2]).

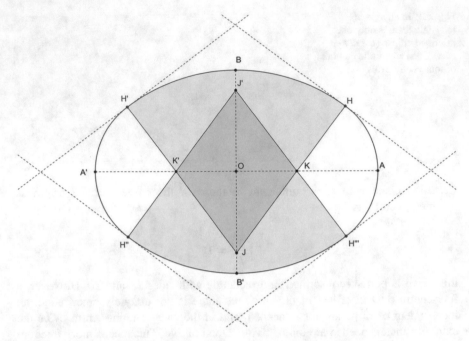

Fig. 2.3 An oval, with its four centres and the tangents at the connecting points

- tangents to the circles in A and B are parallel to the other axis
- J and K belong respectively to the vertical and horizontal axis, with K inside OA and J opposite to B w.r.t. O
- J, K and H are co-linear.

A further well-known property or condition on K is proved by the following argument, inspired by Bosse's construction of an oval in [1], which we will present in Chap. 3. Draw from K the parallel to the chord BH and let F be the intersection with the line OB (see Fig. 2.4). Since $\overline{JH} = \overline{JB}$, we have that $J\widehat{B}H = J\widehat{H}B$, which means that $BFKH$ is an isosceles trapezoid, implying that $\overline{BF} = \overline{KH}$. Since F has to be inside the segment OB, we have the following:

$$\overline{OB} > \overline{FB} = \overline{KH} = \overline{AK} = \overline{AO} - \overline{OK},$$

the first and the last terms yielding that $\overline{OK} > \overline{AO} - \overline{OB}$, which is a condition on point K.

Finally let us assume that for the same K a different second centre J' existed (and thus a different oval). Then—see the bottom of Fig. 2.4—the connecting point would be H', and the same construction as the one just performed would yield a different chord BH' and thus a different point F' inside OB, but this is impossible,

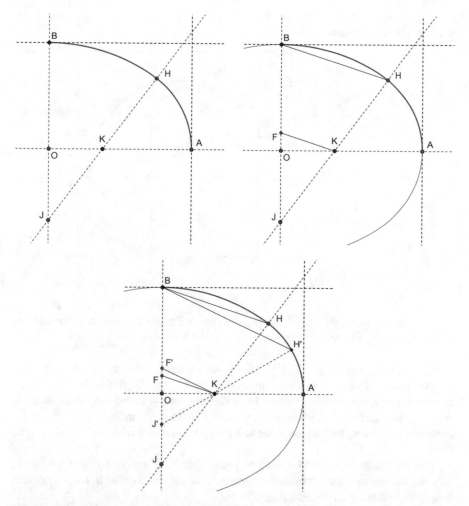

Fig. 2.4 Determination of a condition for K and proving the uniqueness of J and H

because it would still have to be $\overline{F'B} = \overline{KH'} = \overline{AK}$. This means that the choice of a feasible K uniquely determines the quarter oval.

Felice Ragazzo's work on ovals [4], and, later on, the one on polycentric curves [5], made it possible to find new properties and to solve un-tackled construction problems. His 1995 paper presents a property of any connection point and a conjecture on the locus described by all the possible connection points of an oval with given axis measures.

We will present the first property in the form of a theorem.

Theorem 2.1 *In a quarter oval (see Fig. 2.5) the connection point H, the end point B of the smaller axis, and the point P (inside the rectangle inscribing the quarter*

Fig. 2.5 Co-linearity of *B*,
H and *P*, and of *A*, *H* and *Q*

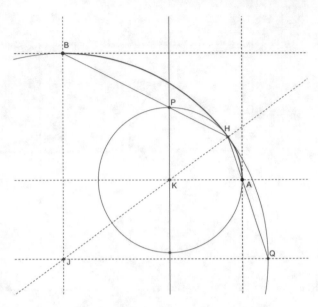

oval) lying on the smaller circle and on the parallel to the smaller axis through the centre K, are co-linear.

Proof Let us connect *B* and *H*. Since *BH* is not a tangent either to the circle through *H* with centre *K* or to the circle through *H* with centre *J*, it will have an extra common point with both. For the bigger circle this is clearly *B*. According to Archimedes' Lemma[2] the intersection with the smaller circle has to lie on the parallel to *BJ* through *K*. Since it has to lie on the same side of *H* and *B*, it has to be *P*. □

The same argument can be used to prove that *H*, *A* and *Q* are also co-linear, *Q* being the intersection between the bigger circle and the parallel to the longer axis through its centre *J* (see again Fig. 2.5).

Ragazzo's conjecture on ovals reduces the problem to *quarter* ovals inside rectangles. It is the following (our translation from the Italian):

The locus of the connection points for the arcs [of a quarter oval [Ed.]] [...] *is the circle, which we will call Connection Locus, defined through the following three points* [points 1, 2 and 3 in Ragazzo's original drawing (Fig. 2.6)]

– *intersection between the half major axis and the half minor side* [of the rectangle inscribing the oval [Ed.]]
– *endpoint of the half minor side transported compass-wise onto the major side*
– *intersection between the half major side and the half minor axis*

---

[2]Given two tangent circles, any secant line through the tangency point intersects the two circles at points whose connecting lines with the corresponding centres are parallel.

Fig. 2.6 Inscribing a quarter oval inside a *rectangle*: Ragazzo's circle through points 1, 2 and 3 is where all possible connection points lie. Figure taken from [4] (copyright of the *Department of Representation and Survey* of the university "La Sapienza" of Rome, with kind permission)

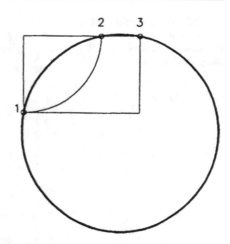

(taken from [4], copyright of the *Department of Representation and Survey* of the university "La Sapienza" of Rome, with kind permission).

In other words, any quarter oval inscribed in a given rectangle will have the corresponding connection point on the above defined Connection Locus, vice versa any point on it (as long as it lies between points 1 and 2, as we will see later) can be chosen as connection point for a quarter oval inscribed in the given rectangle.

We first wrote the proof of this conjecture, then managed to extend it to egg-shapes and to polycentric curves. So it has become in [3] a corollary of one of the theorems contained in [2]. In the framework of those works the chosen tool was that of Euclidean geometry. Another proof based on analytic geometry was presented by Ghione in the appendix to [4]. We present here both proofs, to show the two different possible approaches to the study of ovals, with just a couple additions to the second one to make it equivalent to the first.

Theorem 2.2 *Let OATB be a rectangle with $\overline{BT} > \overline{AT}$. Let S be the point on BT such that $\overline{ST} = \overline{AT}$.*

1. *The necessary and sufficient condition for H to be the connecting point for two arcs of circle with centres inside $B\widehat{T}A$ and tangent in H, one tangent to the line AT in A and the other one tangent to the line TB in B, is for H to belong to the open arc AS of the circle through A, B and S.*
2. *The angle originating at the centre C of such a circle corresponding to the arc AB is a right angle.*
3. *The line OC forms a $\frac{\pi}{4}$ angle with OA*

Proof 1 (Mazzotti 2014) Let us consider Fig. 2.7. We will start from the second part of the theorem. Let C be the centre of the circle through A, B and S; since $\overline{ST} = \overline{AT}$ we have that $T\widehat{S}A = \frac{\pi}{4}$, which implies that $C\widehat{S}A + C\widehat{S}B = A\widehat{S}B = \frac{3}{4}\pi$ and so, since

Fig. 2.7 The quarter *oval* resulting from a choice of *H* on the Connection Locus of the given *rectangle*

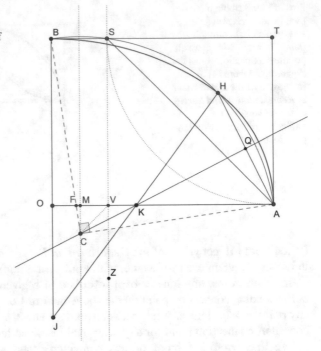

$\widehat{CSA} = \widehat{CAS}$ and $\widehat{CSB} = \widehat{CBS}, \widehat{CAS} + \widehat{CBS} = \dfrac{3}{4}\pi$, looking at the quadrilateral $CASB$ we can deduce that

$$\widehat{ACB} = 2\pi - \frac{3}{4}\pi - \frac{3}{4}\pi = \frac{\pi}{2}. \tag{2.1}$$

For the necessary condition we will show that two generic arcs of circle enjoying the properties stated have their connection point on the described circle. If K on OA is the centre of one of the two, we have (see the remarks preceding the present theorem) $\overline{OA} - \overline{OB} < \overline{OK} < \overline{OA}$.

The connection point belongs to the circle with centre K and radius $\overline{KA}$. We will prove that it is the other intersection between this circle and the one through A, B and S. Let H be this intersection and detect point J lying on both KH and OB; if one proves that $\overline{JH} = \overline{JB}$ then the second circle uniquely determined by A, B and K (see again the remarks preceding the present theorem), having its centre in J and radius equal to $\overline{BJ}$, will be found; the connection point of the two arcs is then clearly H.

Summing up, if H is the (second) intersection of the circle through A, B and S, with the circle with centre K and radius $\overline{KA}$, we have (see Fig. 2.7):

$$Hypothesis: \begin{cases} \overline{BC} = \overline{CS} = \overline{CA} = \overline{CH} \\ \overline{TA} = \overline{TS} \\ \overline{KA} = \overline{KH} \end{cases} \qquad Thesis: \overline{HJ} = \overline{JB}. \qquad (2.2)$$

Equality of the triangles CKA and CKH implies that $K\widehat{H}C = K\widehat{A}C$, which added to $C\widehat{H}B = C\widehat{B}H$ yields

$$K\widehat{H}B = K\widehat{A}C + C\widehat{B}H. \qquad (2.3)$$

Equality (2.1) implies that triangles BOF and FAC are similar, and so that $J\widehat{B}C = K\widehat{A}C$; by substitution of this equality in (2.3) we obtain that $K\widehat{H}B = J\widehat{B}C + C\widehat{B}H = J\widehat{B}H$, hence the thesis.

With regard to the sufficient condition it is enough to choose a point H on arc AS and determine the intersection K of the axis of the segment HA with OA; the circle with centre K and radius $\overline{KH}$ will then pass through A and have AT as tangent; once we show that K exists and is such that

$$\overline{OA} - \overline{OB} < \overline{OK} < \overline{OA} \qquad (2.4)$$

calling J the intersection of KH with OA, the proof that H is the connection point proceeds *formally* as the second part (2.2) of the proof of the necessary condition.

In order to prove (2.4) we will call V the intersection of the parallel to OB through S with OA; we start by pointing out that the equality of triangles AVC and VCS implies that VC divides $Z\widehat{V}O$ in two equal parts, and so that

$$O\widehat{V}C = C\widehat{V}Z = \frac{\pi}{4} \qquad (2.5)$$

Point C is on the axis of BS, thus inside the strip determined by BO and SV. If it were on the same side as H w.r.t. OA, we would have a non-tangled quadrilateral $AOBC$ having the angle in O measuring $\frac{\pi}{2}$ and the one in C measuring $\frac{3}{2}\pi$—the concave version of $A\widehat{C}B = \frac{\pi}{2}$ (notice that (2.1) has been proved independently of K)—which is unacceptable. This implies, since C belongs to KQ (see Fig. 2.7), that point K exists and lies between A and M (intersection of the axis of the segment BS with OA), and so that $\overline{OK} < \overline{OA}$. On the other hand $K \equiv V$ is no possible option, since if $\overline{AV} = \overline{VS}$, we would have $\overline{AK} = \overline{KS}$ and the circle with centre K would have S in common with the circle through A, B and S, i.e. $H \equiv S$, impossible according to our hypotheses. Moreover, it is also impossible that $\overline{OM} < \overline{OK} < \overline{OV}$, otherwise $O\widehat{K}C$ would be an external angle of the triangle KVC, and for a known theorem it would be $O\widehat{K}C > O\widehat{V}C$; this would imply, since $A\widehat{K}Q = O\widehat{K}C$, that $A\widehat{K}Q > O\widehat{V}C$, and thus—observing the triangle QAK—that

$$Q\widehat{A}K = \frac{\pi}{2} - A\widehat{K}Q < \frac{\pi}{2} - O\widehat{V}C \tag{2.6}$$

But $S\widehat{A}K$ is a part of $Q\widehat{A}K$, and so $Q\widehat{A}K > S\widehat{A}K = \frac{\pi}{4}$; comparing this with (2.6) we obtain $\frac{\pi}{2} - O\widehat{V}C > \frac{\pi}{4}$, implying $O\widehat{V}C < \frac{\pi}{4}$, in contrast with (2.5).

Finally, using (2.5), since $OMC = VMC$, we get that $C\widehat{O}A$ also measures $\frac{\pi}{4}$, and the third part has also been proved. □

Proof 2 (Ghione 1995) In order to use Cartesian coordinates, a better reference is that of Fig. 2.8. *ATBD* is the rectangle—with $\overline{BD} < \overline{AD}$—where we are going to inscribe a quarter-oval with symmetry centre D, where A is made to be the origin of an orthogonal coordinate system, AD and AT are chosen as the *x*-axis and the *y*-axis, and is set equal to one. In addition let S be the point on the segment TB such that $\overline{TS} = \overline{TA}$. After having chosen the centre of the smaller circle K on AD, we have the following points:

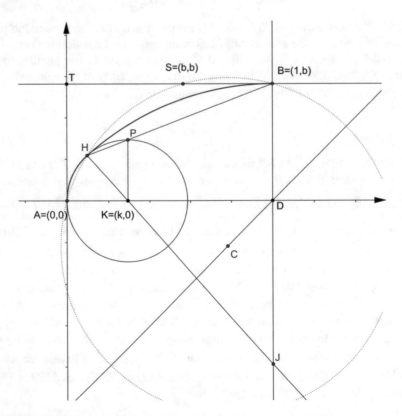

Fig. 2.8 Proving Ragazzo's conjecture by means of analytic geometry (Ghione)

$$A = (0,0), \quad B = (1,b), \quad S = (b,b) \quad \text{and} \quad K = (k,0), \quad \text{where } 0 < k < b < a.$$

TA is tangent to the circle with centre K and radius $\overline{AK}$, which is one of the two circles needed to draw the quarter-oval. We draw the parallel to AT through K and call P the intersection with the circle, then the line through B and P to detect point H as its intersection with the circle. Theorem 2.1 tells us now that H has to be the connecting point for the (only) quarter-oval that can be inscribed in $ATBD$ with K as one of the centres.

The circle has centre $K = (k,0)$ and radius k, thus the equation

$$x^2 + y^2 = 2kx; \tag{2.7}$$

point P has coordinates $P = (k,k)$, and the line through P and B can be described, in parametric form, by the equations

$$\begin{cases} x = k + (1-k) \cdot t \\ y = k + (b-k) \cdot t \end{cases} \tag{2.8}$$

where t varies over real numbers, and P and B are obtained when $t=0$ and $t=1$ respectively. By substitution of these equations in that of the circle (2.7) we obtain the two points corresponding to the values of the parameter $t=0$ (yielding point P) and

$$t = \frac{-2k(b-k)}{(b-k)^2 + (1-k)^2},$$

which corresponds to the connection point H. Substituting these values into (2.8) we get the following coordinates of H:

$$\begin{cases} x = \dfrac{(b-1)^2 k}{(b-k)^2 + (1-k)^2} \\ y = \dfrac{k(b-1)(2k-b-1)}{(b-k)^2 + (1-k)^2} \end{cases}$$

which allow for different positions of H when a different point K is chosen (with $0 < k < b$). To get an equation of the locus of the set of H points for different feasible positions of K, we eliminate k from the last equations: dividing the first one by the second one we get

$$\frac{y}{x} = \frac{2k - (b+1)}{(a-1)}, \text{which implies } k = \frac{(b-1)y + (b+1)x}{2x}.$$

Substitution of the latter into the first in (2.8) yields

$$x^2 + y^2 - (b+1)x - (b-1)y = 0, \tag{2.9}$$

which is clearly the equation of a circle, and it is solved by the coordinates of all three points

$$A = (0,0), B = (1,b) \text{ and } S = (b,b).$$

As k varies continuously in the open interval $]0; b[$ the coordinates of H vary continuously according to formulas (2.8); performing the two limits for $k \to 0$ and for $k \to b$ we get that

$$\lim_{k \to 0} (x,y) = (0,0) \equiv A \text{ and that } \lim_{k \to b} (x,y) = (b,b) \equiv S,$$

which means that *all* points in the open arc AS of the circle through A, B and S are connecting points for some quarter-oval, and that there are no other such points outside it. The preceding argument works thus as proof of both the necessary and the sufficient condition.

From (2.9) we derive the following coordinates of the centre C of the circle through A, B and S:

$$x_C = \frac{(b+1)}{2} \text{ and } y_C = \frac{(b-1)}{2};$$

calculating the slope of the lines CB and CA we get

$$m_{CB} = \frac{\frac{b-1}{2} - b}{\frac{b+1}{2} - 1} = -\frac{b+1}{b-1} \text{ and } m_{CA} = \frac{\frac{b-1}{2} - 0}{\frac{b+1}{2} - 0} = \frac{b-1}{b+1}$$

which implies, since $m_{CB} \cdot m_{CA} = -1$, that $B\widehat{C}A = \frac{\pi}{2}$.

Finally we observe that the coordinates of C, for any value of b belong to the bisector of $A\widehat{D}J$ whose equation is $y = x - 1$. The proof is now complete. $\square$

The fact that CO bisects $J\widehat{O}A$ is another result already conjectured by Ragazzo and proved by Ghione in [4].

Definition *From now on we will call the circle through A, B and S, following Ragazzo's work (1995), Connection Locus (our translation), or simply CL (as in* [3]).

Rosin found the same set of points after Ragazzo did (see [6]) proving its properties, and called it *Locus of continuous tangent joints*. Simpson used it in 1745, to solve a problem posed by Stirling (in [7]) without realizing its full properties.

If $H \equiv S$ the closed convex curve is not an oval anymore, since the arcs of the bigger circle degenerate into segments, and becomes what we call a *running track*. In Sect. 3.4 we will talk about such a shape.

If $H \equiv A$ the curve cannot be considered an oval because the arcs of the small circles disappear and the shape becomes pointed.

We now observe in Fig. 2.7 that QK, the axis of the chord AH of the circle with centre K has to be the same as the axis of the chord AH of the circle with centre C, therefore CQ divides the angle $H\widehat{K}A$ into two equal parts, and does the same with the opposite angle $O\widehat{K}J$. Noting also that (2.5) holds for point V (see again Fig. 2.7), and that $A\widehat{V}T = \dfrac{\pi}{4}$, we have proved the following theorem.

Theorem 2.3 *The centre C of the Connection Locus for a quarter oval is the incentre of triangle OKJ. Moreover the line connecting C with T—the point where the perpendiculars to the axes through A and B meet—forms with OA an angle of $\dfrac{\pi}{4}$.*

We can now sum up all the properties listed and proved in the previous pages, in order to get a complete picture. These are the properties that have been used to derive parameter formulas (in Chap. 4) and to find easier or new constructions (in Chap. 3), especially in the case where the axis lines are not known.

We use for this purpose Fig. 2.9, where the important points in a full oval have been highlighted, although the properties will be listed only for the points needed to draw a quarter-oval.

In a quarter oval with the segments OA and OB as half-axes (with $\overline{OB} < \overline{OA}$), T as fourth vertex of the rectangle having O, A and B as vertices, K as the centre of the smaller arc, J as the centre of the bigger arc and H as the connection point, in addition to the properties listed at the beginning of this section, the following hold:

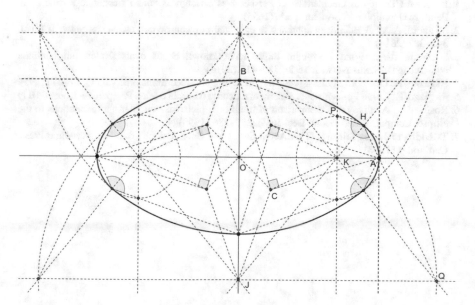

Fig. 2.9 Properties of an oval and its points

- $\overline{AO} - \overline{OB} < \overline{OK} < \overline{OA}$
- *if S is the point on segment BT such that $\overline{TS} = \overline{TA}$, then H belongs to the open arc AS of the circle through A, B and S—the Connection Locus*
- *the centre C of the Connection Locus is the incentre of triangle OKJ*
- *$A\widehat{C}B$ is a right angle, which implies that the radius of the Connection Locus is equal to the side of a square with the same diagonal as that of the inscribing rectangle OATB*
- *if P is the intersection between the parallel to OB through K and the smaller circle, then P, H and B are co-linear*
- *if Q is the intersection between the parallel to OA through J and the bigger circle, then Q, H and A are co-linear*
- *since $A\widehat{H}B$ and $A\widehat{S}B$ cut the same chord AB in the CL circle, we have that*

$$A\widehat{H}B = \frac{3}{4}\pi$$

As we will see in Chap. 3, there are some hidden constraints on the numbers involved, for example, there does not exist a corresponding quarter-oval for any points *A*, *B* and *J* satisfying the above conditions. Moreover other parameters involved, including angles for example, do play a part in the possibility of drawing an oval subject to specific constraints.

References

1. Bosse, A.: Traité des Geometrales et Perspectives Enseignées dans l'Academie Royale de la Peinture et Sculpture. L'Auteur, Paris (1655)
2. Mazzotti, A.A.: A Euclidean approach to eggs and polycentric curves. Nexus Netw. J. **16**(2), 345–387 (2014)
3. Mazzotti, A.A.: What Borromini might have known about ovals. Ruler and compass constructions. Nexus Netw. J. **16**(2), 389–415 (2014)
4. Ragazzo, F.: Geometria delle figure ovoidali. Disegnare idee immagini. **VI**(11), 17–24 (1995)
5. Ragazzo, F.: Curve Policentriche. Sistemi di raccordo tra archi e rette. Prospettive, Roma (2011)
6. Rosin, P.L.: A survey and comparison of traditional piecewise circular approximations to the ellipse. Comput. Aided Geom. Des. **16**(4), 269–286 (1999)
7. Tweddle, I.: James Stirling: 'This About Series and Such Things'. Scottish Academic Press, Cambridge (1988)

Ruler/Compass Constructions of Simple Ovals

We can now show how most ovals can be drawn with ruler and compass if enough parameters are known. And after having mastered the basic ones, one is ready to tackle more complicated problems, such as the two *Frame Problems* or the *Stadium Problem*, which we will discuss in the second part of this chapter. All the following constructions have been made with freeware Geogebra and most of them are linked through the website www.mazzottiangelo.eu/en/pcc.asp, as described further. The much used *Connection Locus*—the *CL*—has just been defined in Chap. 2. All constructions in this chapter are *general purpose* constructions, in the sense that any combination of parameters, constrained within some values, can be chosen. Further combinations of parameters then those illustrated here are listed in the Appendix, for the reader to try out and find the corresponding constructions. Selected oval forms will then be presented in Chap. 6.

Sources for this chapter include our recent work [7] and Dotto's paper [5]. New issues include two of the basic constructions which were missing in [7] (nos. 10 and 11) and the formulation and solution to the (direct and inverse) *Frame Problem*.

Issues left open are the "missing" ruler/compass Constructions 8 and 13, and some kind of proof of the suggested limitations for the parameters in the constructions listed in Sect. 3.2.

All the constructions are meant for a quarter oval, and then simply reflected w.r.t. the two axes. When a ruler/compass construction of a whole oval is to be performed, slight changes may have to be made in order to minimize the number of passages required.

Considering only a quarter oval we will use the following parameter. If O is the centre of symmetry, let

- $a = \overline{OA}$ the length of half the major axis
- $b = \overline{OB}$ the length of half the minor axis
- $k = \overline{OK}$ the distance from O of the centre of the smaller circle
- $j = \overline{OJ}$ the distance from O of the centre of the bigger circle

© Springer International Publishing AG 2017
A.A. Mazzotti, *All Sides to an Oval*, DOI 10.1007/978-3-319-39375-9_3

- h the distance of the connecting point H from OB
- m the distance of the connecting point H from OA
- r_1 the length of the radius $\overline{AK}$
- r_2 the length of the radius $\overline{BJ}$
- $\beta = A\widehat{K}H$ the angle formed by the line of the centres and the major axis
- $p = \frac{\overline{OB}}{\overline{OA}}$ the ratio of the two axes

3.1   Ovals with Given Symmetry Axis Lines[1]

First of all we will deal with ovals with known symmetry axis lines with O as symmetry centre. The formulas connecting the different parameters, corresponding to the 23 constructions described in this section are derived in the next chapter, numbered in the same way.

The conjecture stated in [7] was:

> *The first six parameters are independent of one another, and any choice of three of them—constrained within certain values—will determine a unique oval.*

This conjecture has proved *not* to be true, because two of the four unsolved cases in [7] (out of 20) were solved and one of them delivered values for the parameters involved which allow for *two* ovals (Construction 11). Moreover Cases 8 and 13 are still partially unsolved. The main six parameters are independent just the same and they form an efficient way of representing a four-centre oval and classifying constructions on the basis of given data.

Table 3.1 lists the 20 different choices of three out of the first six of the above parameters, and the number of the corresponding Case/Construction, for further easier references. Constructions 8 and 13 are the ones missing, while Constructions 9 and 10 are completely new. After these constructions others follow, including Constructions 23b and 72, which have been presented for the first time in [7].

In the following constructions some parameters are subject to constraints, but feasible values always exist. Exact limitations are justified—via analytic calculations—in Chap. 4.

Sometimes more than one construction is displayed in order to present the work of different scientists. Constructions very similar to others already shown are left to the reader.

Construction 1—given a, b and k, with $0 < a - b < k < a$

1a. Bosse's construction (Fig. 3.1). Bosse's construction (1655) may be the first published general purpose construction (see [1]). It is the same as the one published by Tosca in 1712 (see [9]):

---

[1]Most of this section has already been published in the Nexus Network Journal (see [7])

Table 3.1 The oval construction numbers according to the given parameters, with known symmetry axis lines

Construction no.	a	b	k	h	j	m
1	●	●	●			
2	●	●		●		
3	●	●			●	
4	●	●				●
5	●		●	●		
6	●		●		●	
7	●		●			●
8	●			●	●	
9	●			●		●
10	●				●	●
11		●	●	●		
12		●	●		●	
13		●	●			●
14		●		●	●	
15		●		●		●
16		●			●	●
17			●	●	●	
18			●	●		●
19			●		●	●
20				●	●	●

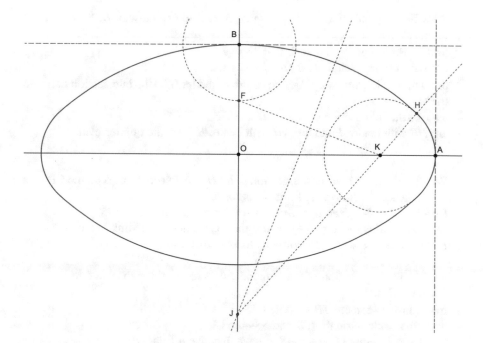

Fig. 3.1 Construction 1a—Bosse's method

- find F on segment BO such that $\overline{BF} = \overline{KA}$
- draw the axis of segment FK
- let J be the intersection of this line with BO
- draw the line KJ and let H be the intersection with the circle having centre K and radius $\overline{KA}$
- arc HB with centre J and arc AH with centre K form the quarter-oval.

1b. Ragazzo's construction (Fig. 3.2). Theorem 2.1 has proved what Ragazzo has stated in [8]. This is his way of constructing on oval given a, b and k.

- let P be the intersection of the parallel to line BO from K and the circle with centre K and radius $\overline{AK}$
- draw the line BP and let H be the intersection with the drawn circle
- let J be the intersection of KH with the line BO
- arc HB with centre J and arc AH with centre K form the quarter-oval.

1c. Yet another method (Fig. 3.3) uses Ragazzo's CL and is thus mathematically supported by Theorem 2.2; see also the link to the Geogebra animation video on www.mazzottiangelo.eu/en/pcc.asp):

- draw the perpendiculars to OA through A, and to OB through B; let T be their intersection
- find S inside segment TB such that $\overline{TS} = \overline{TA}$
- draw the circle through A, B and S —the CL
- draw the circle with radius $\overline{AK}$ and centre K and let H be the intersection between the two
- let J be the intersection of lines KH and OB
- arc HB with centre J and arc AH with centre K form the quarter-oval.

We already know that the connection point H is between A and S on the CL (see Fig. 2.7), hence the following limitations for h.

Construction 2—given a, b and h, with $0 < a - b < h < a$

In this construction (Fig. 3.4) we use the CL again (see the link to the Geogebra animation video on www.mazzottiangelo.eu/en/pcc.asp):

- let T be the intersection of the perpendiculars to OA through A, and to
- OB through B;
- find S inside segment TB such that $\overline{TS} = \overline{TA}$
- draw the circle through A, B and S —the CL
- find the only point H on arc AS to have distance h from OB
- draw the axis of segment AH and let K be the intersection with OA

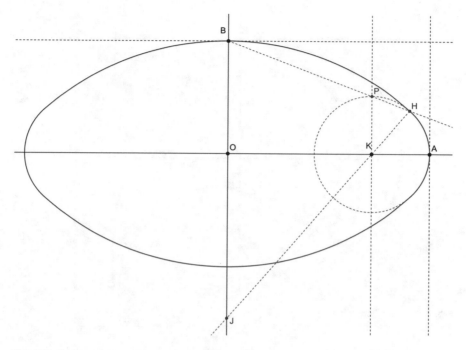

Fig. 3.2 Construction 1b—Ragazzo's method

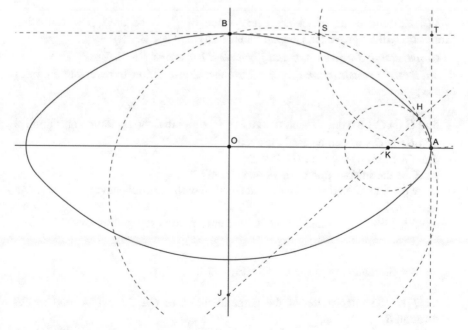

Fig. 3.3 Construction 1c

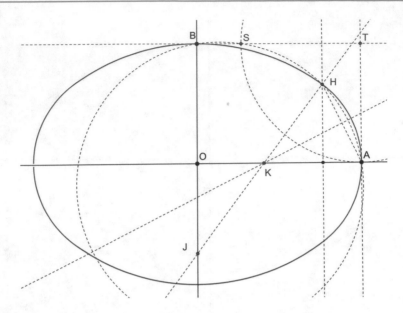

Fig. 3.4 Construction 2

– let *J* be the intersection of lines *KH* and *OB*
– arc *HB* with centre *J* and arc *AH* with centre *K* form the quarter-oval.

When *a* and *b* are given not every value of *j* can be used. The following limitation will be proved in Chap. 4 when studying Case 3.

Construction 3—given a, b and j, with $0 < b < a$ **and** $j > \frac{a^2 - b^2}{2b}$

3a. Bosse's construction (Fig. 3.5). We use Construction 1a reversing the roles of the parameters

– find *G* on *OA* beyond *O* such that $\overline{AG} = \overline{JB}$ (note that the limitation on *j* implies that $\overline{BJ} > \overline{OA}$ as can be easily proved)
– draw the axis of segment *GJ*
– let *K* be the intersection of this line with *AO*
– draw the line *KJ* and let *H* be the intersection with the circle having centre *K* and radius $\overline{KA}$
– arc *HB* with centre *J* and arc *AH* with centre *K* form the quarter-oval

3b. An alternative is to use the CL (Fig. 3.6).

– let *T* be the intersection of the perpendiculars to *OA* through *A*, and to *OB* through *B*
– find *S* inside segment *TB* such that $\overline{TS} = \overline{TA}$

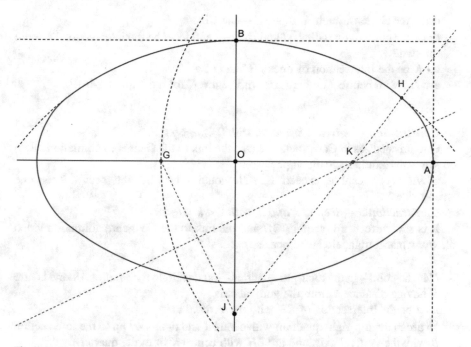

Fig. 3.5 Construction 3a

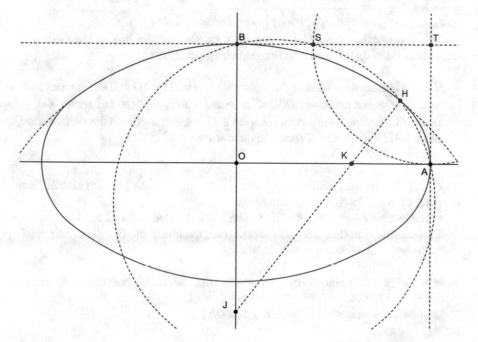

Fig. 3.6 Construction 3b

- draw the circle through A, B and S —the CL
- draw the circle with radius $\overline{BJ}$ and centre J and let H be the intersection between the two
- let K be the intersection of lines JH and OA
- arc HB with centre J and arc AH with centre K form the quarter-oval.

Construction 4—given a,b and m, with $0 < m < b < a$

One proceeds as in Construction 2 (see the link to the Geogebra animation video on www.mazzottiangelo.eu/en/pcc.asp).

Obviously k cannot exceed h, KH would otherwise intersect OB on the wrong side.

Construction 5—given a, k and h, with $0 < k < h < a$

It is straightforward (see Fig. 3.7 and the link to the Geogebra animation video on www.mazzottiangelo.eu/en/pcc.asp):

- H is the point—in the top right quadrant—on the circle with radius $\overline{AK}$ and centre K having distance h from the vertical axis
- let J be the intersection of KH with the vertical axis
- an arc in the top-right quadrant with centre J and radius $\overline{JH}$ up to the intersection B with the vertical axis, and arc AH with centre K form the quarter-oval.

Since b is not known, any positive value for j and any value for k not exceeding a are feasible.

Construction 6—given a, k and j, with $0 < k < a$ and $j > 0$

This construction (Fig. 3.8) is also very simple (see the link to the Geogebra animation video on www.mazzottiangelo.eu/en/pcc.asp):

- H is the intersection—inside the right angle formed by AO and the vertical axis, containing K but not J—of JK and the circle with radius $\overline{AK}$ and centre K
- an arc with centre J and radius $\overline{JH}$ up to the intersection B with the vertical axis, and arc AH with centre K form the quarter-oval.

Point H lies on the circle with radius $a - k$, therefore it can't be that distant from OA. Hence the following limitation for m.

Construction 7—given a, k and m, with $0 < k < a$ and $0 < m < a - k$

This construction (Fig. 3.9) is as follows (see the link to the Geogebra animation video on www.mazzottiangelo.eu/en/pcc.asp):

- H is the nearest point to A on the circle with radius $\overline{AK}$ and centre K having distance m to OA
- let J be the intersection of lines KH and OB

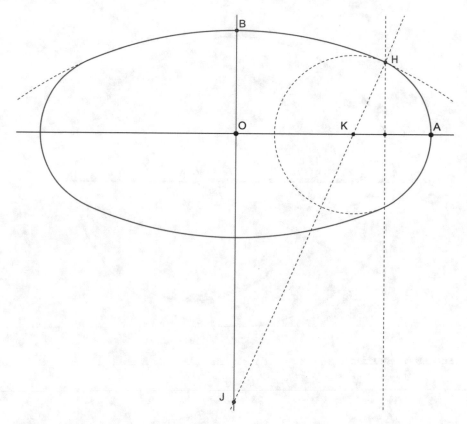

Fig. 3.7 Construction 5

– an arc with centre J and radius $\overline{JH}$ up to the intersection B with the vertical axis, and arc AH with centre K form the quarter-oval

Construction 8—given a, h and j, with $0 < h < a$ ***and*** $j > 0$

This construction has not been found yet. As will be seen in the next chapter (Case 8) only an implicit solution for one of the missing parameters could be derived.

The following condition on parameter m will be explained in Chap. 4, Case 9.

Construction 9—given a, h ***and*** m, ***with*** $a,h > 0$ ***and*** $0 < a - h < m < \sqrt{a^2 - h^2}$

This construction (Fig. 3.10) is as follows (see the link to the Geogebra animation video on www.mazzottiangelo.eu/en/pcc.asp):

– K is the intersection between the axis of AH and OA
– let J be the intersection of lines KH and OB

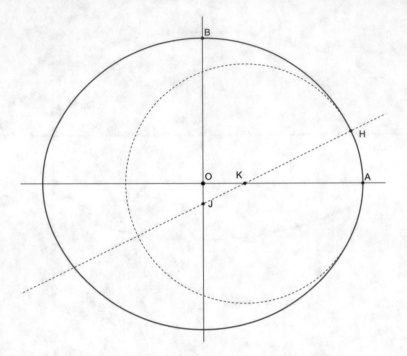

Fig. 3.8 Construction 6

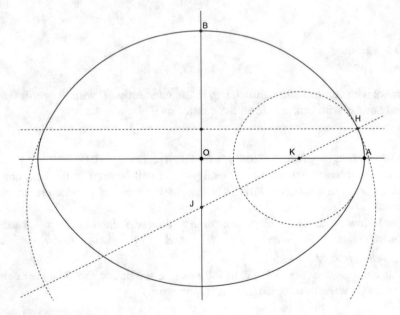

Fig. 3.9 Construction 7

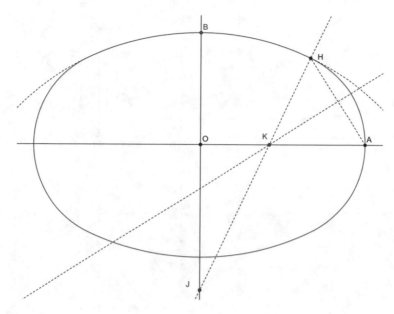

Fig. 3.10   Construction 9

- an arc with centre J and radius $\overline{JH}$ up to the intersection B with the vertical axis, and arc AH with centre K form the quarter-oval.

The three following constructions, nos. 10, 11a and 11b, are totally new ones, and they have been deduced from the formulas derived for Case 10 and Case 11 in Chap. 4. With additional work one should come up with simpler ones.

Construction 10—given a, j and m, with $j > 0$ and $0 < m < b$

This lengthy construction (Fig. 3.11) is as follows (see the link to the Geogebra animation video on www.mazzottiangelo.eu/en/pcc.asp):

- draw the top half circle having segment AO as diameter and find the intersection C with the circle with centre O and radius $\overline{OM} = m$
- on the line OC find a point R beyond C such that $\overline{CR} = \overline{OJ}$
- draw the triangle ACR and the height CH_1 w.r.t. AR
- build a right-angled triangle CLD having $CL//OA$, $CL = CH_1$ as cathetus and the hypothenuse CD lying on the line OC
- find E on line AC beyond C such that $\overline{CE} = \overline{CD}$
- build a right-angled triangle EAG having segments EA and AG as catheti where $\overline{AG} = \overline{AC}$
- let N inside EC be such that $\overline{EN} = \overline{DL}$

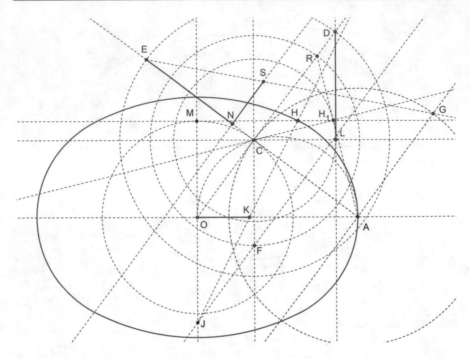

Fig. 3.11 Construction 10

- finally let S be the intersection between the perpendicular to EC through N and EG
- K is now found inside segment OA such that $\overline{OK} = \overline{NS}$
- H is found as the intersection between JK and the parallel to OA through M
- an arc with centre J and radius $\overline{JH}$ up to the intersection B with the vertical axis, and arc AH with centre K form the quarter-oval.

The following conditions on the parameters come from the formulas which will be presented in Chap. 4—Case 11. A further oval can be found if $h > b$ also holds (see Construction 11b).

Construction 11a—given b, k and h, with $b > 0$ and $0 < k < h < \sqrt{b^2 + k^2}$.

This construction (Fig. 3.12) is also rather complicated (see the link to the Geogebra animation video on www.mazzottiangelo.eu/en/pcc.asp). It finds one of the *two* ovals which match any feasible combination of the above parameters, corresponding to the first solution in (4.11):

- let M be the point on KO beyond K such that $\overline{OM} = h$
- draw the bottom half-circle having segment MO as diameter and find the intersection P with the circle with centre O and radius $\overline{OB}$

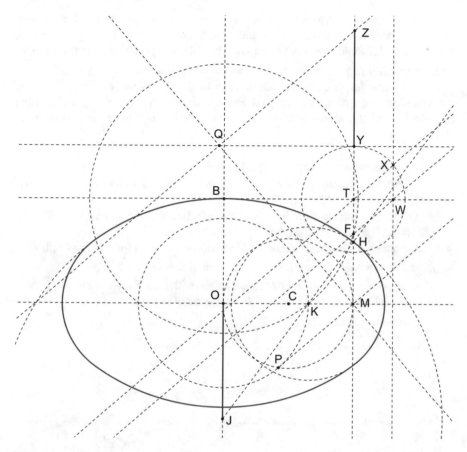

Fig. 3.12 Construction 11a

- let F be the intersection between the circle with centre B and radius $\overline{BK}$ and the parallel to OB through M
- let T be the intersection between the perpendicular to OB through B and MF
- connect P with M and draw the parallels to this line through F and T; let W be the intersection between the first parallel and BT, and let X be the intersection between the second one and the parallel to OB through W
- let Y be the intersection between MT and the circle with centre T and radius $\overline{TX}$, on the *opposite* side of M w.r.t. T
- let Q be the intersection between the parallel to OM through Y and the orthogonal line to MP through M
- draw from Q the perpendicular to QM and let Z be the intersection of it with MY
- point J is the point on line BO beyond O such that $\overline{OJ} = \overline{YZ}$.
- H is the point where JK and FM meet; an arc with centre J and radius $\overline{JH}$ from H to B, and arc AH with centre K form the quarter-oval.

What makes the described situation special is that Construction 11a delivers one of the *two* possible ovals—given *b*, *k* and *h*—because, as we will justify in Chap. 4, if $h > b$, an additional one can be drawn. The following construction shows how.

Construction 11b—given b, k and h, with $0 < b < h$ ***and*** $0 < k < h < \sqrt{b^2 + k^2}$.

This construction (Fig. 3.13) (see the link to the Geogebra animation video on www.mazzottiangelo.eu/en/pcc.asp) delivers a second oval matching any feasible combination of the above parameters, corresponding to the second solution in (4.11):

- follow Construction 11a until point X is found
- let Y' be the intersection between MT and the circle with centre T and radius $\overline{TX}$ on the *same* side of M w.r.t. T
- let Q' be the intersection between the parallel to OM through Y' and the orthogonal line to MP through M
- draw from Q' the perpendicular to $Q'M$ and let Z' be its intersection with MY'
- point J' is the point on line BO beyond O such that $\overline{OJ'} = \overline{Y'Z'}$.
- H' is the point where $J'K$ and FM meet; an arc with centre J' and radius $\overline{J'H'}$ from H' to B, and arc AH' with centre K form the quarter-oval.

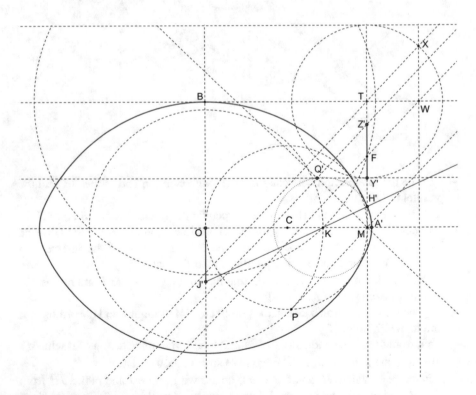

Fig. 3.13 Construction 11b

Check Case 12 in Chap. 4 for the following limitations on k.

Construction 12—given b, k and j with $b, j > 0$ **and** $0 < k < \sqrt{b^2 + 2bj}$
One proceeds as in Construction 6 (see the link to the Geogebra animation video on www.mazzottiangelo.eu/en/pcc.asp).

Construction 13—given b, k and m, with $0 < m < b$ **and** $k > 0$
This construction has not been found yet. As will be seen in the next chapter (Case 13) only an implicit solution for one of the missing parameters could be derived.

Check Case 14 in Chap. 4 for the following limitations on h.

Construction 14—given b, h and j, with $b, j > 0$ **and** $0 < h < \sqrt{b^2 + 2bj}$
One proceeds as in Construction 5 (see the link to the Geogebra animation video on www.mazzottiangelo.eu/en/pcc.asp).

The following limitations are needed in order for J to be on the opposite side of B w.r.t. O.

Construction 15—given b, h and m, with $h > 0$ **and** $0 < m < b < \sqrt{h^2 + m^2}$
One proceeds as in Construction 9 (see the link to the Geogebra animation video on www.mazzottiangelo.eu/en/pcc.asp).

Construction 16—given b, j and m, with $j > 0$ **and** $0 < m < b$
We proceed as in Construction 7 (see the link to the Geogebra animation video on www.mazzottiangelo.eu/en/pcc.asp).

Construction 17—given k, h and j, with $j > 0$ **and** $0 < k < h$
The construction (Fig. 3.14) is as follows (see the link to the Geogebra animation video on www.mazzottiangelo.eu/en/pcc.asp):

– H is the point $K\widehat{O}B$ on line JK with distance h from OJ, opposite to J w.r.t. K
– an arc with centre J and radius $\overline{JH}$ up to the intersection B with the vertical axis, and an arc with centre K and radius $\overline{KH}$ up to the intersection A with the horizontal axis form the quarter-oval.

Construction 18—given k, h and m, with $m > 0$ **and** $0 < k < h$
The construction (Fig. 3.15) is as follows (see the link to the Geogebra animation video on www.mazzottiangelo.eu/en/pcc.asp):

– let J be the intersection of KH with the vertical axis
– an arc with centre J and radius $\overline{JH}$ up to the intersection B with the vertical axis, and an arc with centre K and radius $\overline{KH}$ up to the intersection A with the horizontal axis form the quarter-oval.

Construction 19—given k, j and m, with $k, j, m > 0$
One proceeds as in Construction 17 (see the link to the Geogebra animation video on www.mazzottiangelo.eu/en/pcc.asp).

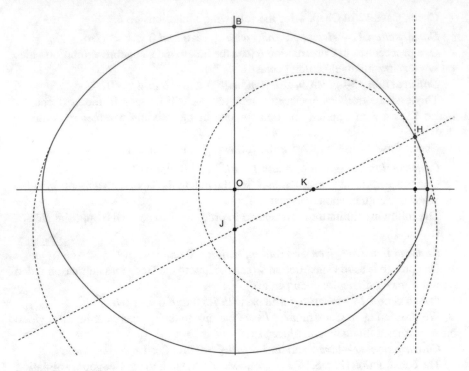

Fig. 3.14 Construction 17

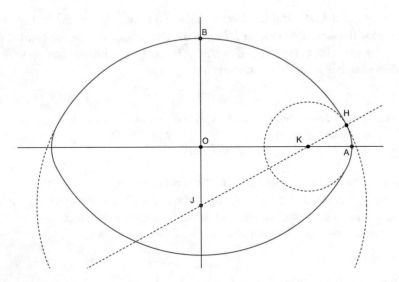

Fig. 3.15 Construction 18

Construction 20—given h, j and m, with $h, j, m > 0$

One proceeds as in Construction 18 (see the link to the Geogebra animation video on www.mazzottiangelo.eu/en/pcc.asp).

Although the six parameters seem to share the same dignity—hence the conjecture at the beginning of this chapter—when a mixed combination is given (b and k with either h or m, or a and j with either h or m) the solutions become complicated and/or give way to exceptions.

If one considers any possible choice of three among the ten parameters listed at the beginning of this chapter, the construction problems amount to a total of 116—four less than the number of combinations because in four cases the parameters are not independent of one another (e.g. a, k and r_1). We have given numbers to these combinations, for future references, and the list is in the Appendix of this book. For many of them one can automatically use one of the above constructions, but this is not always the case.

Three of these constructions are presented here, numbered as in the Appendix[2]. The first one is presented in Huygens' version and in a similar version that the author came up with using the CL. The second one is a new construction, already presented in [7], while the third one is a very simple construction involving the two radii.

The limitation for β will be discussed in Chap. 4, Case 23.

Construction 23—given a, b and β, with $0 < b < a$ *and* $2arctg\frac{a}{b} - \frac{\pi}{2} < \beta < \frac{\pi}{2}$

23a. Huygens' construction (Fig. 3.16). This construction, as explained in [5], was originally intended for an angle $\beta = \frac{\pi}{3}$, but the extension to (nearly) any angle β is automatic: in this paper Dotto cites as sources [2, 3] and [4] (where a table is dedicated to different constructions of polycentric ovals, including arches with 5, 7 or 11 arcs):

- draw an angle equal to β onto OA, with vertex O.
- draw a circle with centre O and radius $\overline{OA}$ and name D and C the intersections with OB and the second side of the β angle
- draw the segment DC and its parallel from B, and call H the intersection of this line with the segment AC
- the parallel to OC through H is the line where K and J can be found as intersections with the two axes; the construction ends as usual.

Triangle ODC is built to be isosceles, which is why BHJ, similar to ODC, also has the two needed equal sides JB and JH. In addition to this AHK, being similar to OCA, is also isosceles, and $\overline{AK} = \overline{KH}$. The arcs can thus be drawn and they share a common tangent because K, H and J are co-linear.

---

[2]This numbering differs from that in [7]

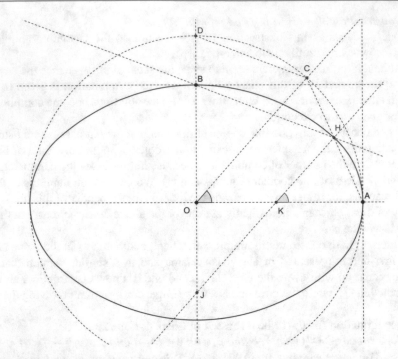

Fig. 3.16 Construction 23a—Huygens' method

23b. It is similar to Huygens' construction (Fig. 3.17), in the sense that an isosceles triangle is drawn, but the point H is found using the CL (see the link to the Geogebra animation video on www.mazzottiangelo.eu/en/pcc.asp):

- draw a half line with origin O forming the given angle β
- with centre O draw an arc with radius b intersecting the half line in D
- let C be the vertex of the right-angled isosceles triangle with hypotenuse AB (the one on the same side of O w.r.t. AB)
- draw the CL with centre C and radius $\overline{CB}$, letting H be the intersection with BD
- draw from H the parallel to OD; intersections with the axes are the centres K and J of the arcs forming the quarter-oval.

Construction 72 is a very recent construction (see [7]), drawing the quarter-oval given the centres of the arcs and the ratio p between the half axes (Fig. 3.18). Theorem 2.3 implies that point C is the intersection between the bisectors of $O\widehat{K}J$ and $K\widehat{O}J$, and that T is on the line through C forming an angle of $\frac{\pi}{4}$ with OK. The limitations for the value of p have to do with the existence of a point T in the right quadrant, and they will be explained in Chap. 4 (Case 72).

Construction 72—given k, j and p, with $k,j>0$ and $\dfrac{(j+k)\left(j-k+\sqrt{j^2+k^2}\right)}{2jk} < p < 1$

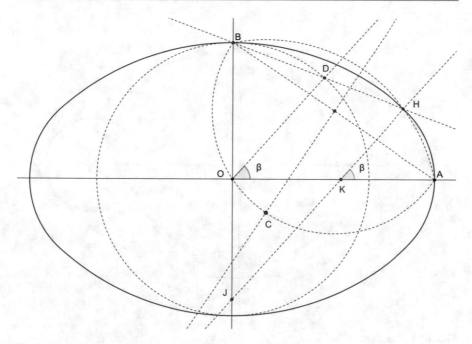

Fig. 3.17 Construction 23b

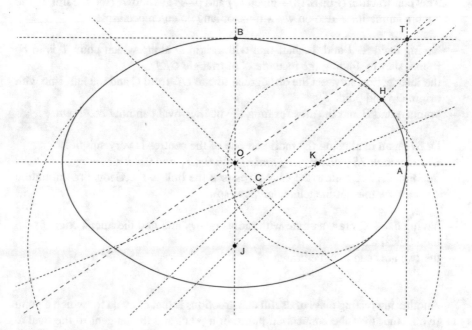

Fig. 3.18 Construction 72

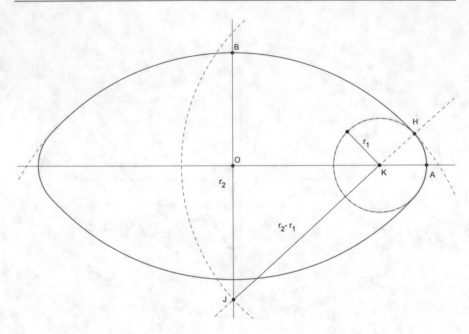

Fig. 3.19   Construction 78

This construction (Fig. 3.18)—given k, j and p—is as follows (see the link to the Geogebra animation video on www.mazzottiangelo.eu/en/pcc.asp):

– having fixed K, J and the half line with origin O along which point T is to be found, start by finding the incentre C of triangle OKJ
– the intersection between the orthogonal line to OC from C and the half line with origin O yield point T
– points A and B and the arcs forming the quarter-oval can now be drawn.

Drawing an oval using the radii and one of the centres is very simple.
Construction 78—given r_1, r_2 ***and*** k***, with*** $0 < r_1$ ***and*** $0 < k < r_2 - r_1$
See Fig. 3.19 to follow the easy steps (see the link to the Geogebra animation video on www.mazzottiangelo.eu/en/pcc.asp):

– having fixed K, draw a circle with radius $r_2 - r_1$ and find the intersection J with the vertical axis
– the two arcs can now be drawn

Another interesting class of useful constructions is that of ovals for which a point is given, other than the connection point, or a point and the tangent to the oval at that point, together with some other parameter. Since they correspond to inscribing or circumscribing ovals, we have left some of these to Sect. 3.3.

3.2   Ovals with Unknown Axis Lines[3]

We will now consider the problem of drawing an oval on a plane without any given symmetry axes; material from this section is taken from [7]. We now have three more degrees of freedom and so a much bigger variety of combinations. We will nevertheless show with a few examples that parameters that look independent aren't always so.

The CL and the constructions used in the first part of this chapter are the tools to tackle this problem. Limitations conjectured for the parameters involved are maybe the most interesting part. We consider constructions U21, U22 and U23 to be the most interesting ones.

Fig. 3.20 is going to be our new reference.

An oval in a plane is determined by six independent parameters, although limitations do occur. We will again try to be systematic and give numbers to the different problems/solutions.

In what follows A and A' are opposite vertices on the longer axis, B and B' on the shorter one, J and J' centres of the arcs with longer radii, K and K' of the ones with shorter radii, and finally H, H', H'' and H''' the connection points. We will again use C to indicate the centre of a CL.

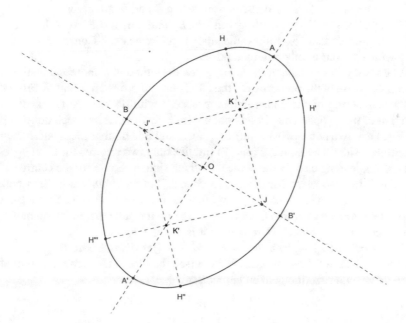

Fig. 3.20 A generic oval in a plane

---

[3]Most of this section has already been published in the Nexus Network Journal (see [7])

Given on a plane A and A' opposite vertices of an oval—corresponding to four parameters—the symmetry centre O will be their midpoint, and lines AA' and its orthogonal through O will be the symmetry axes. One continues now by choosing any two parameters other than a from Table 3.1 using the corresponding constructions already listed. The ten possible choices will be numbered from U1 to U10[4] following the order in which they appear in Table 3.1.

Constructions U1 to U10—given A, A' and any other two independent parameters other than a—follow the above guidelines once O and the two axes have been drawn. As an example we will draw Construction U7 (see the link to the Geogebra animation video on www.mazzottiangelo.eu/en/pcc.asp):

Construction U7—given A, A', k and m with the limitations of Construction 7.
See Fig. 3.21 to follow the construction:

– O is the midpoint of AA', from which we draw the orthogonal to it
– a parallel distant m from AA', and the circle with a centre point K—on AA' distant k from O—meet in H.
– we find now J intersecting HK and the already drawn perpendicular to AA', and draw the complete oval as usual, using symmetry properties

The same ideas allow for constructions when B and B' are given.

Constructions U11 to U20—given B, B' and any other two independent parameters other than b—follow the above guidelines for Constructions U1 to U10, with a similar numbering method.

When two non-opposite vertices are given the situation is more interesting, and we can make use of the properties of the CL listed and proved in Chap. 2. From now on the numbering of the constructions proceeds with no specific criterion, other than that of their appearance in this book (similarly to what has been done in [7].

Given any two non-opposite vertices, say A and B, a double CL is automatically determined, since Theorem 2.2 (part 2) implies only two choices for C on opposite sides of AB, leaving us only the choice of which side to choose for the centre C. We then have still two more parameters to choose, but we *cannot* use them both to choose H because the connection point must be on an arc of the CL. So we can still choose the symmetry centre O, which has to be on the half-circle with diameter AB already containing point R. Here is how it is done.

Construction U21—given any A, B and then (feasible) H and O
See Fig. 3.22 to follow the construction (see the link to the Geogebra animation video on www.mazzottiangelo.eu/en/pcc.asp):

---

[4]Again the numbering differs from that in [7]: the letter U has been added and 99 has been subtracted (e.g. Oval 105 has become Construction U6)

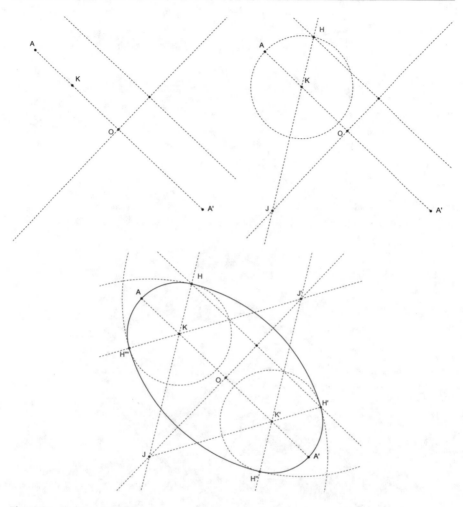

Fig. 3.21 Construction U7

- let C and C_1 be the vertices of the two right-angled isosceles triangles with hypotenuse AB, centres of a CL; choose one of them, say C, and choose any point H on the arc AB with centre C
- let P be the intersections of the parallel to BH through A with the circle having diameter AB, and Q the intersection of the parallel to AH through B with the semi-circle; choose O, the symmetry centre, on the arc PQ.
- one way of drawing our arcs is now, for example, to find J as the intersection of the axis of BH with OB, and then K as the intersection of JH with OA.

Limits for O on the arc PQ are based on construction evidence, *not* on mathematical proof. Choosing the centre of symmetry between C and Q exchanges the

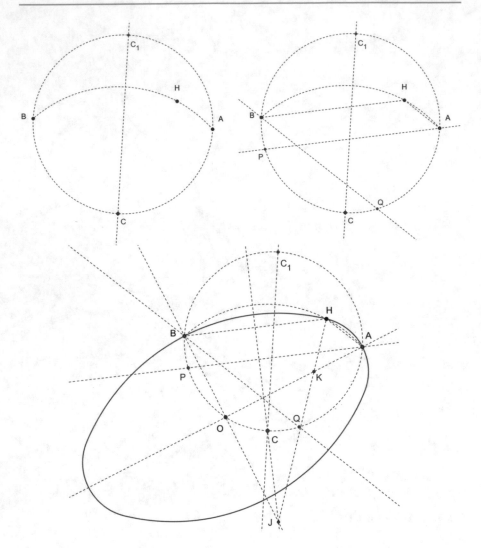

Fig. 3.22 Construction U21

roles of A and B. If, on the other hand, one wants to choose O first, then H will have to be taken inside AS after having found S (see Fig. 2.7).

It is also possible to choose either K or J, after having chosen A and B, each counting for two extra parameters, since they can be freely chosen inside two dimensional areas, as we have learned—although *not* proved—via construction evidence. We believe that when feasible values for O in Construction U21 are proved, then feasible values for K in U22 can be derived. See the following constructions U22 and U23.

Construction U22—given any A, B and then (feasible) K

In Fig. 3.23 note the supposed area of feasibility for K bordered in green (see the link to the Geogebra animation video on www.mazzottiangelo.eu/en/pcc.asp):

- let C and C_1 be the vertices of the two right-angled isosceles triangles with hypotenuse AB; choose one of them, say C, draw the CL arc between A and B, and then mirror the smaller of the two CA arcs—of the circle with diameter AB— w.r.t. the line CA

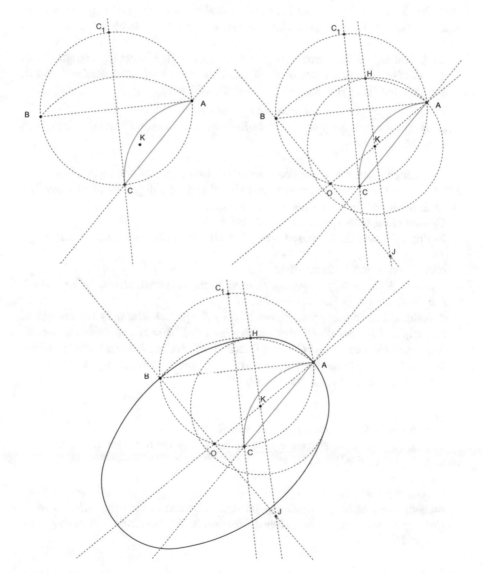

Fig. 3.23 Construction U22

- choose any point K *inside the just determined segment corresponding to the chord CA*
- AK is now the symmetry axis, and the connecting point H is the intersection between the CL with centre C and the circle with centre K and radius AK
- determine O as the intersection between AK and the circle with diameter AB, and then J as the intersection of OB—the other axis—with line HK and so on.

Construction U23—given any A, B and then (feasible) J
In Fig. 3.24 note the supposed area of feasibility for J bordered in green (see the link to the Geogebra animation video on www.mazzottiangelo.eu/en/pcc.asp):

- let C and C_1 be the vertices of the two right-angled isosceles triangles with hypotenuse AB; choose one of them, say C, draw the CL arc between A and B, and then the line CB, which forms an angle of $\frac{\pi}{4}$ with CC_1
- choose any point J *inside the angle opposite to $B\widehat{C}C_1$*
- JB is now the symmetry axis, O is the intersection between JB and the circle with diameter AB, and so on.

Another choice for a starting pair is that of two centres of non-equal circles, say J and K. Constructions are very straightforward and we will just outline the starting steps. Limitations for the points are conjectured.
Construction U24—given any J, K and A
In Fig. 3.25 note the supposed area of feasibility for A bordered in green:

- choose any J and K, then either
- *A inside the open half-plane determined by the perpendicular to JK through K not containing J, but not on line JK,* or
- *A inside the open half-plane determined by the perpendicular to JK through K, containing J but not on JK, and outside the circle with centre J and radius JK*
- draw the axis KA and find O as the intersection between KA and the perpendicular to it from J; if O lies between A and K what you have is actually A', and to get the *real A* you need to mirror it w.r.t. O.

Construction U25—given any J, K and B
In Fig. 3.26 note the supposed area of feasibility for B bordered in green; the construction is equivalent to U24:

- choose any J and K
- then choose *B inside the open half-plane determined by the perpendicular to JK through J containing K, excluding JK and excluding the circle with centre J and radius JK*

Fig. 3.24 Construction U23

– draw the axis *BJ* and find *O* as the intersection between *BJ* and the perpendicular to it from *K*; in half the cases you get a mirrored version of Fig. 3.21.

Constructions U26, U27 and U28 - given any J, K, (feasible) H and either O or A or B

Fig. 3.25 Initial steps for
Construction U24

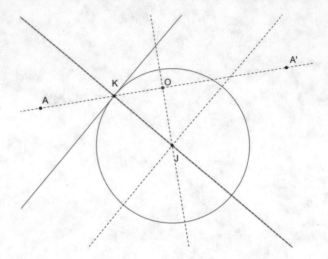

Fig. 3.26 Initial steps for
Construction U25

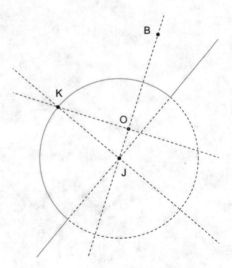

In Fig. 3.27 see how these constructions should start (see the link to the
Geogebra animation video on www.mazzottiangelo.eu/en/pcc.asp):

- choose any *J* and K
- choose *H* on *JK* beyond *K*
- choose either
- *O on the circle with diameterJK excluding J and K*, or
- *A on the open half circle YHZ excluding H*, or
- *B on the open half-circle WHX excluding H*.

Fig. 3.27 Initial steps for Constructions U26, U27 and U28

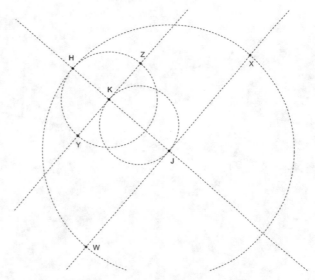

As a last example we will consider a given tangent and its tangent point, in addition to points A and K. Bosse's construction is also used here.

Construction U29—given any K, A, an extra point C supposed to lie on the same half oval as A, and the tangent to the oval in C

It is as follows (see Fig. 3.28; also see the link to the Geogebra animation video on www.mazzottiangelo.eu/en/pcc.asp):

- choose any A and K
- draw the circle with centre K and radius $\overline{AK}$; let D be the point opposite to A
- draw the tangent through D and the two other tangents perpendicular to the first one
- choose C *inside one of the right angles formed by the first and second, or the first and third tangent, in both cases opposite to the one containing the circle* (see Fig. 3.28)
- draw from C the tangent to the first circle (the one on A's side), and the parallel to KA
- draw the desired tangent through C *between the two lines just drawn* (see Fig. 3.28)
- take a segment $CF = KA$ on the normal line through C (on the side of the first circle) find J intersecting the normal line from C and the axis of segment KF, etc.

Point C can be also chosen on the other side of the tangent through D, but in that case the possible choices for the tangent through C are more limited. We believe that the above limitations for C are somewhat conservative, but further investigations need to be made.

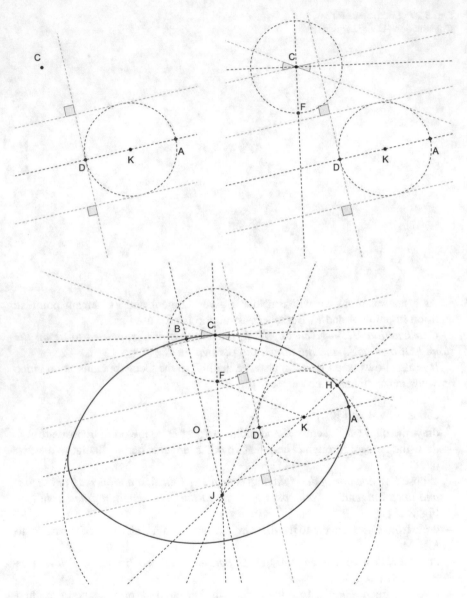

Fig. 3.28 Construction U29

A lot more combinations of the parameters listed can be chosen, and constructions found, using the new tool of the CL and its properties along with the previously known properties and constructions.

3.3 Inscribing and Circumscribing Ovals: The Frame Problem

Inscribing an oval in a rectangle is as easy as circumscribing an oval around a rhombus. The axis lengths a and b are automatically given and one is free to choose an extra parameter among the remaining seven (excluding p) thus using for example one among Constructions 1, 2, 3, 4, 21, 22 and 23. An infinite number of ovals can be drawn (see Fig. 3.29).

It is more interesting to study how many different ovals can be inscribed in a rhombus or circumscribed around a rectangle, or constrained to fill the gap between two rectangles.

An oval inscribed in a rhombus has to be tangent to all of the sides, but due to symmetry it is enough to study the case of a single side. First of all it is important to understand which of the two diagonals of the rhombus is met first by the perpendicular to the side from the chosen point of tangency. In the special case of the centre of the rhombus O lying on this perpendicular, the only inscribed oval is... a circle, and no other choices can be made. In all other cases the diagonal which is first met by the perpendicular is going to be the line containing the longer axis. Let this be the horizontal diagonal (Fig. 3.30). One chooses the corresponding intersection as the centre of the smaller circle K, and the endpoint A of the longer axis is now determined by the radius $\overline{KP}$. An extra choice can now be made as long as the limitations for the chosen parameter are met, in addition to some other (e.g. $b < \overline{OD}$ or $j > \overline{OR}$). Figure 3.31 illustrates a set of possible solutions. Note that for one of these we can choose $H \equiv P$.

There is, however, another possibility: the intersection with the *vertical axis* of the *same* perpendicular can be chosen as centre J of the bigger circle, yielding a whole new set of ∞^1 ovals (see Fig. 3.32), again with some restrictions on the parameters.

Let P now be a point on side DG such that the perpendicular to the side meets the vertical diagonal first. In this case we can either use this point as K, or the intersection with the horizontal axis as J. In the first case we get an infinite number of possible ovals among which the three shown in Fig. 3.33. In the second case we get another group, which the ones selected for Fig. 3.34 belong to.

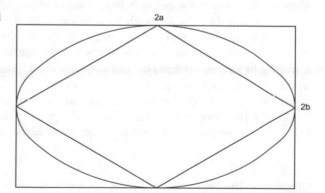

Fig. 3.29 An oval inscribed in a rectangle and circumscribing a rhombus

Fig. 3.30 Choosing the
centre of a small circle for an
oval inscribed inside a
rhombus

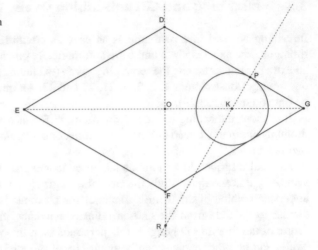

Fig. 3.31 A set of inscribed
ovals corresponding to the
choice made in Fig. 3.30

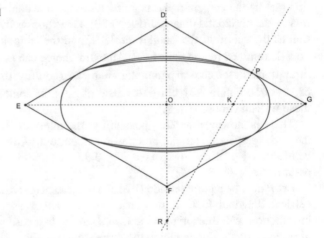

Circumscribing a rectangle with an oval doesn't require any particular skill,
since one can draw any segment from a vertex to one of the symmetry axes and then
proceed in exactly the same ways as with the ovals inscribed in a rhombus (see
Fig. 3.35 for a set of possible solutions among the ∞^2).

The situation becomes much more interesting when an oval is required which is
inscribed in a rectangle and circumscribes a second rectangle inside the first one,
sharing the same symmetry axes. When the outer rectangle is given we call this the
frame problem. The construction is straightforward once the right conditions are
met for the vertices of the inner rectangle.[5] The proof of the limitations for the inner
rectangle will be given at the end of Chap. 4, via analytic calculations (Case 117).

---

[5]Since a point does not come in the form of a parameter, we number this construction 117.

Fig. 3.32 Another set of
infinite ovals inscribed in the
same rhombus given the same
tangency point on one of the
sides

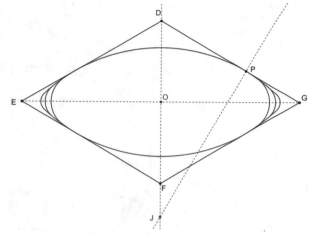

Fig. 3.33 The case when the
perpendicular meets the
vertical diagonal first:
choosing K on it

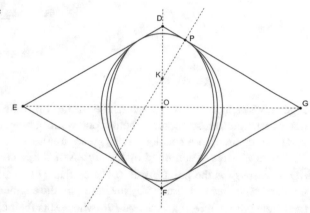

Fig. 3.34 The case when the
perpendicular meets the
vertical diagonal first:
choosing J on the other one

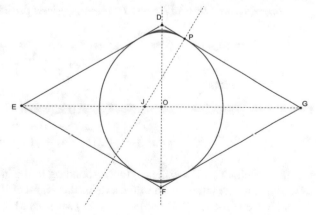

Fig. 3.35 Some solutions to the problem of circumscribing a rectangle with an oval

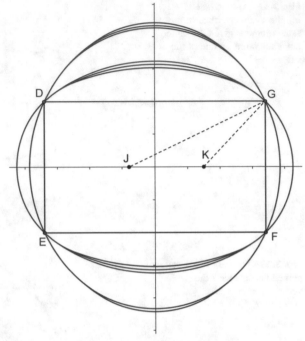

Figure 3.36 shows the two areas where in the top right hand side a vertex of the inner rectangle can be chosen, whether one wants a bigger circle to run through this vertex (blue area) or a smaller circle (green area).

The green area is delimited by the arc AS of the circle with centre Q and radius $\overline{AQ}$—where Q is the point inside OA such that $\overline{QA} = \overline{OB}$—and the arc AS of the CL. The blue area is delimited by the latter, by the segment BS and by the arc AB of the circle with centre P and radius $\overline{PB}$, where P is the point on BO beyond O such that $\overline{BP} = \overline{PA}$.

Construction 117 (the Frame Problem)—given positive a, b and a point $(\bar{x}; \bar{y})$ such that

$$\textit{either a)} \quad \begin{cases} \bar{x}^2 + \bar{y}^2 + \bar{y}\left(\dfrac{a^2 - b^2}{b}\right) - a^2 > 0 \\ \bar{x}^2 + \bar{y}^2 + \bar{x}(b - a) + \bar{y}(a - b) - ab < 0 \\ 0 < \bar{y} < b \end{cases}$$

$$\textit{or b)} \quad \begin{cases} \bar{x}^2 + \bar{y}^2 + 2\bar{x}(b - a) + a^2 - 2ab < 0 \\ \bar{x}^2 + \bar{y}^2 + \bar{x}(b - a) + \bar{y}(a - b) - ab > 0 \end{cases}$$

If our point $X(\bar{x}; \bar{y})$ satisfies a) (corresponding to the blue area in Fig. 3.37), then J is found as the intersection of the axis of the segment BX with OB and we can go on as in Construction 3. If our point $Y(\bar{x}; \bar{y})$ satisfies b) (corresponding to the green area in Fig. 3.38), then K is found as the intersection of the axis of the segment AY with OA and we can go on as in Construction 1.

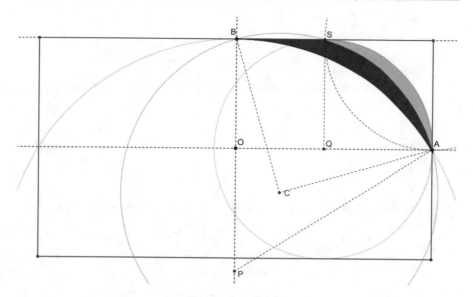

Fig. 3.36 Areas where a vertex can be chosen to solve *the frame problem*

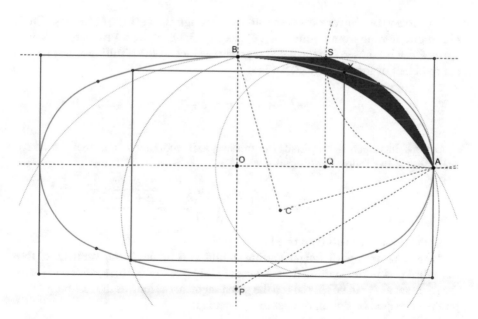

Fig. 3.37 A solution to the *frame problem* choosing a vertex in the blue section

When the inner rectangle is given we have the *inverse frame problem*, corresponding to Constructions 118a and 118b.

Construction 118 (the Inverse Frame Problem)

This is about finding feasible axis measures for an oval circumscribing a given rectangle. 118a.

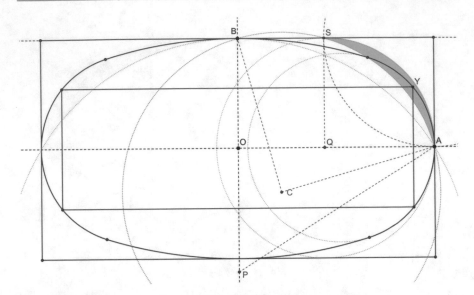

Fig. 3.38 A solution to the *frame problem* choosing a vertex in the green section

The arcs with *shorter* radius should run through the vertices of the inscribed rectangle. Starting from a vertex $P(\bar{x}; \bar{y})$ (see Fig. 3.39), a point A to share the same arc as P can be chosen (on the green segment) as long as the resulting K lies inside the right half of the rectangle. That is if

$$\sqrt{\bar{x}^2 + \bar{y}^2} < a < \bar{x} + \bar{y}.$$

Once K has been determined, the missing axis endpoint B can then be taken according to the following limitations:

$$a - k < b < \frac{a(\bar{x} - k) - k\left(\bar{x} - a + \sqrt{(a - \bar{x})(a + \bar{x} - 2k)}\right)}{\bar{x} - k}.$$

Both inequalities will be proved in Chap. 4.

***118b.* The arcs with *longer* radius should run through the vertices of the inscribed rectangle.** Starting from a vertex $P(\bar{x}; \bar{y})$ (see Fig. 3.40), a point B to share the same arc as P can be chosen (on the green segment) as long as the resulting J lies on the other side of B with respect to O. That is if

$$\bar{y} < b < \sqrt{\bar{x}^2 + \bar{y}^2},$$

as will be proved in Chap. 4.

Once J has been determined, the missing axis endpoint A can then be taken according to the following limitations:

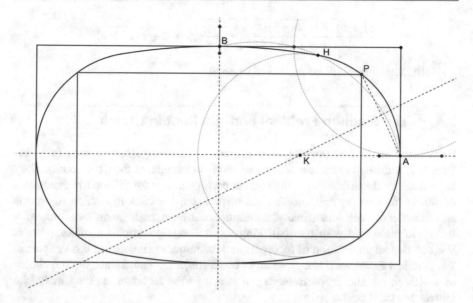

Fig. 3.39 A solution to the *inverse frame problem* choosing first *A* then *B* on the green segments

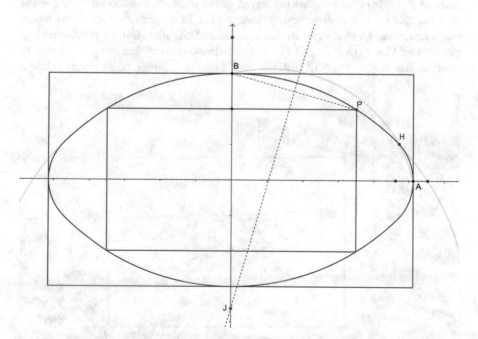

Fig. 3.40 A solution to the *inverse frame problem* choosing *B* and then *A* on the green segments

$$\frac{\bar{y}(j+b) + \sqrt{(b-\bar{y})(b+\bar{y}+2j)}}{j+\bar{y}} < a < \sqrt{b(2j+b)}.$$

Both inequalities will be proved in Chap. 4.

3.4 The Stadium Problem and the Running Track

Polycentric ovals were popular when it came to planning amphitheatres (see also Sect. 8.2, dedicated to the use of 8-centre ovals in the plan of the Colosseum). They were used for the inside arena, the outside perimeter, and for the intermediate rows of sitting (or standing) places. The use of this shape is back in fashion for sports arenas, after a time when many stadiums have been built using the form of a rectangle prolonged with two half circles, what one could call a *running track*. We can find an application of theorems and techniques presented in this chapter in *The Stadium Problem*. It is a problem of ovals subject to constraints, but the choice of which forms and proportions to use is left to the architect, as long as he/she knows what can be drawn and how.

Suppose we have a well-defined area where a stadium should be built. Suppose then, as it is often the case, that the layout of the stadium is supposed to be a set of cased simple ovals with common symmetry axes. In geometrical terms, this would mean inscribing an oval in an outer rectangle and circumscribing a smaller rectangle with a different oval (Fig. 3.41). Between those two one can imagine a series of intermediate ovals. We will analyse four different classes of solutions without

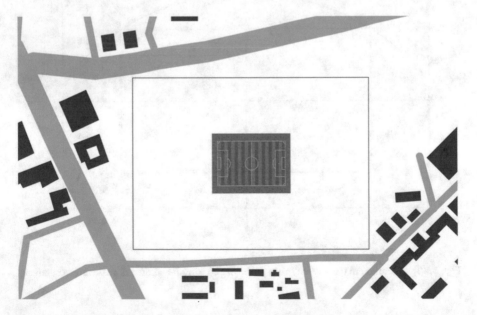

Fig. 3.41 The area where the layout of a stadium should be planned

going into the details of the constructions (see the link to the Geogebra animation video on www.mazzottiangelo.eu/en/pcc.asp).

One way of doing so (see Fig. 3.42) is to start with one of the infinite ovals that can be inscribed in the outer rectangle, and then draw the only circumscribing oval similar to the first one. Three intermediate ovals are then added as an example of the structure of a stadium layout. These ovals all have the same shape, but arcs of different ovals have different centres.

Another way is to start with an inscribed oval and then use the same set of centres to draw the circumscribed one (Fig. 3.43). The intermediate ones will also have the same set of centres. These are called concentric ovals, and Sect. 4.4 is devoted to them. This was the most common choice in roman amphitheatres, as we will show in the case study of the Colosseum, in Chap. 8, although with 8-centre ovals. But two pairs of concentric ovals probably also appear in Borromini's project for San Carlo alle Quattro Fontane as suggested at the end of Sect. 7.2.4.

A third possibility is again to start with an oval inscribed in the outer rectangle and then to project the outer rectangle and create a solvable *Frame Problem* (see previous section) around the inner rectangle, in order to find a circumscribing oval. See Fig. 3.44 for the complete drawing.

Starting from the inside is also possible. Choose one of the ∞^1 ovals circumscribing the inner rectangle having a horizontal longer axis (see Fig. 3.35), and then choose a line where the connection points should lie (Fig. 3.45).

As already said, the *running track* has come a bit out of fashion as a stadium shape, but it is often still used as a starting shape from the inside, since many stadiums are used for athletics competitions. By *running track* we mean a shape

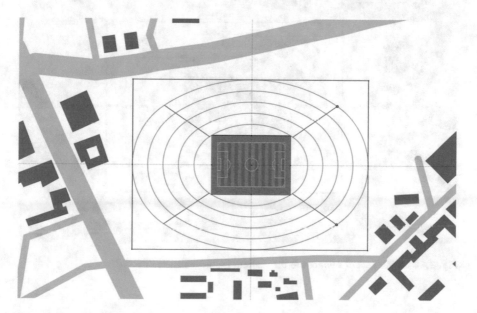

Fig. 3.42 A first set of solutions to the stadium problem

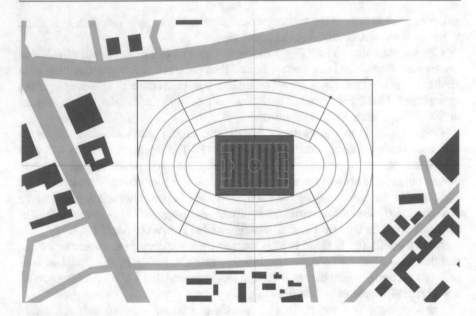

Fig. 3.43 A second set of solutions to the stadium problem

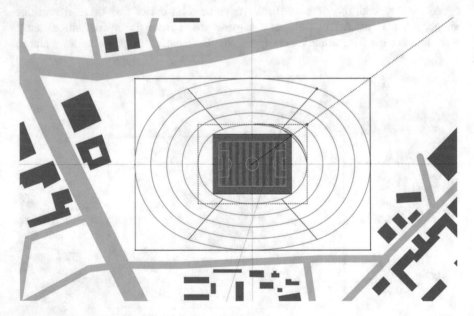

Fig. 3.44 A third set of solutions to the stadium problem

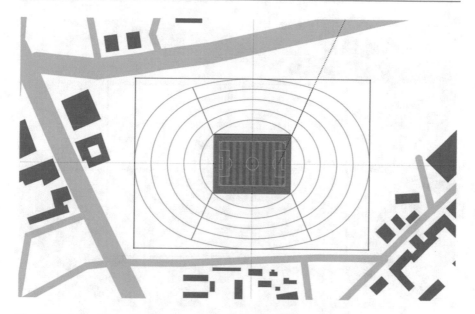

Fig. 3.45 A fourth set of solutions to the stadium problem

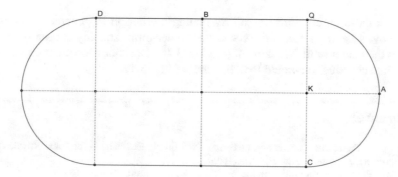

Fig. 3.46 A *running track*

which is made of two segments and two half circles (see Fig. 3.46), so it is *not* an oval in the sense used here. Once the rectangle is fixed (∞^2 choices), the arcs have to be taken with one of the two axes (usually the shorter one) as diameter. But one can consider it as a borderline case of an oval, since the segments can be considered as arcs of a circle having its centres at infinity. We can go back to Fig. 3.4 and imagine what the oval would look like if H were pushed up to point S on the Connection Locus.

By *regular running track* we mean a running track such that $\overline{DQ}$ is equal to arc QC, as is often the case in sports events. This yields $\overline{DQ} = \frac{\pi}{2}\overline{QC}$.

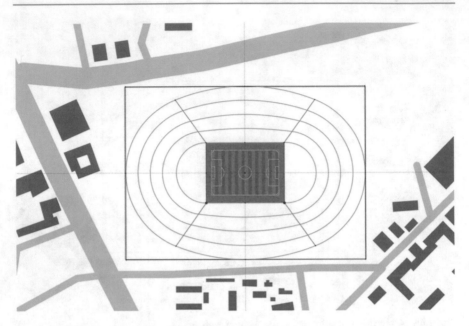

Fig. 3.47   Nested ovals surrounding a running track

Using these observation we can try finding a criterion to generate a series of nested ovals having a *running track* as their inner limit, and any chosen inscribed oval as their outer limit. Thinking of Q in Fig. 3.46 as a connection point for an oval, a possible construction would look like that of Fig. 3.47.

References

1. Bosse, A.: Traité des Geometrales et Perspectives Enseignées dans l'Academie Royale de la Peinture et Sculpture. L'Auteur, Paris (1655)
2. Curioni, G.: Geometria pratica applicata all'arte del costruttore. Torino (1877)
3. Docci, M., Migliari, R.: Architettura e geometria nel Colosseo di Roma. In: Atti del Convegno (ed.) Matematica e architettura. Metodi analitici, metodi geometrici e rappresentazioni in architettura, Università di Firenze, Fac. Architettura, Firenze 13–15 gennaio 2000, pp. 13–24. Alinea Editrice, Firenze (2001)
4. Donghi, D.: Il manuale dell'architetto. Torino (1925)
5. Dotto, E.: Note sulle costruzioni degli ovali a quattro centri. Vecchie e nuove costruzioni dell'ovale. Disegnare Idee Immagini. **XII**(23), 7–14 (2001)
6. Mazzotti, A.A.: A Euclidean approach to eggs and polycentric curves. Nexus Netw. J. **16**(2), 345–387 (2014)
7. Mazzotti, A.A.: What Borromini might have known about ovals. Ruler and compass constructions. Nexus Netw. J. **16**(2), 389–415 (2014)
8. Ragazzo, F.: Geometria delle figure ovoidali. Disegnare idee immagini. **VI**(11), 17–24 (1995)
9. Tosca i Masco, T.V.: Compendio matemàtico: Que comprehende Arquitectura civil, montea, y canteria, arquitecura militar, pirotechnia. Tomo V. Por Antonio Bordazar, Valencia (1712)

Parameter Formulas for Simple Ovals and Applications

4

Many tools for drawing a polycentric oval subject to geometrical or aesthetical constraints have been presented in Chap. 3. In this chapter we derive formulas which on the other hand allow to calculate all important parameters when the value of three independent ones is known, as well as the limitations the given parameters are subject to. These formulas allow for a deeper insight in the properties of any oval, as will be shown on the chosen forms of Chap. 6. Cases included are numbered consistently with the construction numbers of Chap. 3. Also included is the proof of the formulas yielding the solution to the frame problem presented in Sect. 3.3 as well as formulas for the length of an oval and for the area surrounded by it. The final section is devoted to concentric ovals and their properties.

The best framework to use to derive formulas for the different parameters involved in an oval is the Cartesian plane.

4.1 Parameter Formulas for Simple Ovals

The material presented here has mostly been deduced, by means of solving simple analytic geometry problems, from the constructions presented in Chap. 3, but the converse is also true: formulas derived from other formulas have been the basis for elaborate constructions which would not otherwise have been found. It is also true that a direct application of the constructions may lead to rather complicated formulas, so sometimes a different method needs to be used. Many of these formulas have also been derived and used by Rosin in [5], Dotto in [1], Ragazzo in [4], López Mozo in [2] and many others.

Again in this section two cases are left partly unsolved; these have solutions which at the moment are only in implicit form, but the most important one is the exception of Case 11, where quite surprisingly two sets of solutions have been found, proving the conjecture at the beginning of Chap. 3 (first stated in [3]) to be false. The different cases are numbered in the same way as the corresponding constructions in Chap. 3.

© Springer International Publishing AG 2017
A.A. Mazzotti, *All Sides to an Oval*, DOI 10.1007/978-3-319-39375-9_4

The procedures to obtain such formulas are in some cases explained and in some other cases left to the reader, when they are similar to the ones previously described.

Again the ten parameters involved are the ones listed at the beginning of Chap. 3:

$$a, \ b, \ k, \ j, \ h, \ m, \ r_1, \ r_2, \ \beta \, \text{and} \, p,$$

defined as (see the left hand picture in Fig. 4.1)

- $a = \overline{OA}$ the length of half the major axis
- $b = \overline{OB}$ the length of half the minor axis
- $k = \overline{OK}$ the distance from O of the centre of the smaller circle
- $j = \overline{OJ}$ the distance from O of the centre of the bigger circle
- h the distance of the connecting point H from OB
- m the distance of the connecting point H from OA
- r_1 the length of the radius $\overline{AK}$
- r_2 the length of the radius $\overline{BJ}$
- $\beta = A\widehat{K}H$ the angle formed by the line of the centres and the major axis
- $p = \frac{\overline{OB}}{\overline{OA}}$ the ratio of the two axes

Before showing how to obtain three of the first six parameters once the remaining three are known, we will write down the elementary formulas to calculate the last four parameters when a, b, k and j are known.

For r_1, r_2, β and p we have:

$$r_1 = a - k, \quad r_2 = b + j, \quad \beta = arctg\frac{j}{k}, \quad p = \frac{b}{a} = tg\gamma. \qquad (4.1)$$

The most convenient way to fix a coordinate system for deriving the formulas connecting the different parameters is that of Fig. 4.1 (right hand picture).

A formula which we will use is the equation of the CL (see the definition after the two proofs of Theorem 2.2), whose centre C (see for example Theorem 2.2) lies at the intersection of the axis of the segment BS with the bisector of the fourth quadrant, which yields the coordinates $C = \left(\frac{a-b}{2}, \frac{b-a}{2}\right)$. The radius is equal to the distance

$$\overline{AC} = \sqrt{\left(\frac{a-b}{2} - a\right)^2 + \left(\frac{b-a}{2}\right)^2} = \frac{\sqrt{2}}{2}\sqrt{a^2 + b^2}.$$

Simple calculations deliver the following equation of the CL:

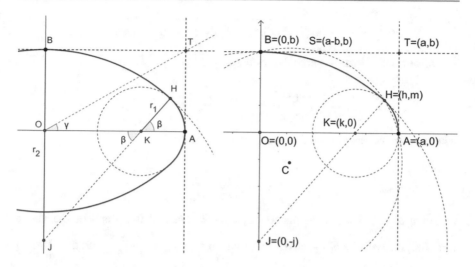

Fig. 4.1 Parameters of an oval and the coordinates of the points determining a quarter-oval

$$x^2 + y^2 + x(b - a) + y(a - b) - ab = 0. \tag{4.2}$$

Another useful equation linking parameters a, b, j and k is the following one:

$$a - b = j + k - \sqrt{j^2 + k^2}, \tag{4.3}$$

which comes from the equalities (see Fig. 4.1) $\overline{OB} + \overline{OJ} = \overline{JK} + \overline{KH} = \overline{JK} + \overline{KA}$, the last and first term yielding $b + j = \sqrt{j^2 + k^2} + (a - k)$ and so equation (4.3). From this equation another useful one can be derived, as Rosin did in [5],

$$k = \frac{j - \frac{a-b}{2}}{\frac{j}{a-b} - 1}$$

which eventually leads to

$$k = \frac{(2j - a + b)(a - b)}{2(j - a + b)}.$$

Before listing the different cases corresponding to the constructions presented in Chap. 3, we derive the formula for r_2 as a function of r_1, for any fixed a and b.

Deriving j and k from the first two equalities in (4.1) and substituting them into (4.3) yields

$$b = a - (r_2 - b) - (a - r_1) + \sqrt{r_2^2 + b^2 - 2br_2 + a^2 + r_1^2 - 2ar_1},$$

which evolves into

$$r_2 - r_1 = \sqrt{r_2^2 + b^2 - 2br_2 + a^2 + r_1^2 - 2ar_1},$$

and after taking squares—note that $r_2 > r_1$—into

$$ar_1 + br_2 - r_1r'2 = a^2 + b^2$$
$$2,$$

where r_2 can be seen as the following function of r_1:

$$r_2(r_1) = \frac{ar_1 - \frac{a^2+b^2}{2}}{r_1 - b}.$$

Since $a - b < k$ we have that $0 < r_1 = a - k < b$, and the graph of the above function, for $r_1 \in \,]0; b[$ —considering that $\lim_{r_1 \to 0^+} r_2(r_1) = \frac{a^2+b^2}{2b}$ and that $\lim_{r_1 \to b^-} r_2$ $(r_1) = +\infty$—is the hyperbola segment represented in Fig. 4.2.

The following cases have the same numbering as the constructions in Chap. 3, so Table 3.1 can still be used as reference for the first 20, and the Appendix for the remaining cases.

Case 1—Given a, b and k, where $a - b < k < a$ (Fig. 4.3).

Define point F as the point inside the segment OB such that $\overline{KA} = \overline{BF}$. Point $F = (0, b - (a - k))$ is the same point described at the beginning of Chap. 2 (see also the top right-hand side of Fig. 2.4), because the oval determined by a, b and k is unique, therefore J is the vertex of an isosceles triangle with base FK. Point J is then the intersection between the y-axis and the axis of FK. Since the slope of the FK line is

$$m_{FK} = \frac{y_F - y_K}{x_F - x_K} = \frac{b - a + k}{-k} = \frac{a - b - k}{k}$$

and the midpoint of segment FK is $M = \left(\frac{k}{2}, \frac{b-a+k}{2}\right)$, we have that the wanted line has slope $-\frac{k}{a-b-k}$ and equation

$$y - \frac{b - a + k}{2} = -\frac{k}{a - b - k}\left(x - \frac{k}{2}\right),$$

intersection with the line $x = 0$, yield for J the coordinates

$$J = \left(0, \frac{(a - b)(a - b - 2k)}{2(b + k - a)}\right).$$

Remembering that $j = \overline{OJ}$ we get the value of the parameter j

Fig. 4.2 The graph of r_2 as a function of r_1, considering acceptable values

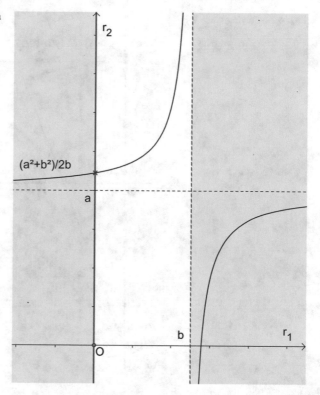

$$j = \frac{(a-b)(2k+b-a)}{2(b+k-a)}. \tag{4.4}$$

Once j is also known there are different ways to calculate h and m. To avoid tedious calculation it is probably better to find H as the intersection between KJ and PB, where P is the point defined at the end of Chap. 2 (see also Fig. 2.9). Since the coordinates of P are $P = (k, a-k)$ we get, for lines PB and KJ respectively, the equations

$$\frac{y-b}{a-k-b} = \frac{x}{k} \quad \text{and} \quad \frac{y}{-j} = \frac{x-k}{-k},$$

solving the corresponding system we get the coordinates of the connection point H:

$$H = \left(\frac{k(b+j)}{b-a+k+j}, \frac{j(a-k)}{b-a+k+j} \right),$$

which are the values of h and m

Fig. 4.3 Solving Case 1

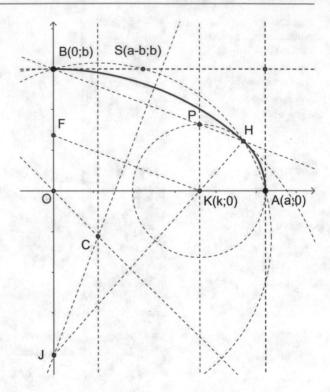

$$h = \frac{k(b+j)}{b-a+k+j} \quad \text{and} \quad m = \frac{j(a-k)}{b-a+k+j} \tag{4.5}$$

Case 2—Given a, b and h, where $0 < a - b < h < a$ (Fig. 4.4).

We just need to find the positive solution y to the system formed by the CL (4.2) and $x = h$. After substituting we get

$$y^2 + \dot{y}(a-b) + h^2 - h(a-b) - ab = 0,$$

and the positive solution is given by

$$m = \frac{(b-a) + \sqrt{(a+b)^2 - 4h(h-a+b)}}{2}.$$

Having calculated m, to find k we can look for the intersection with the x-axis of the axis of segment AH whose equation is

$$y - \frac{m}{2} = \frac{a-h}{m}\left(x - \frac{a+h}{2}\right);$$

and this yields for k the value

Fig. 4.4 Solving Case 2

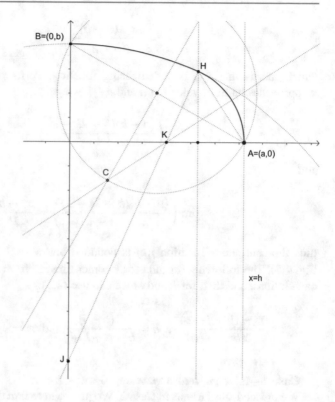

$$k = \frac{a^2 - m^2 - h^2}{2(a-h)}. \tag{4.6}$$

Finally intersecting KH (whose equation is $y(h - k) = m(x - k)$) with the y-axis one gets

$$j - \frac{mk}{h-k}. \tag{4.7}$$

Case 3—Given a, b **and** j, **where** $0 < b < a$ **and** $j > \frac{a^2-b^2}{2b}$.

To justify the above limitations for j let us consider, for fixed a and b, how j varies as a function of k. Let us consider formula (4.4). We perform the derivative with respect to k:

$$\frac{d}{dk}\left(\frac{(a-b)(b+2k-a)}{2(b+k-a)}\right) = \frac{(a-b)}{2} \cdot \frac{2(b+k-a) - (b+2k-a)}{(b+k-a)^2} =$$

$$= -\frac{(a-b)^2}{2(b+k-a)^2} < 0$$

which means that $j(k)$ is a decreasing function. Performing now the limits as k approaches its limit values of a and $a-b$ we get

$$\lim_{k \to a-b} \left(\frac{(a-b)(b+2k-a)}{2(b+k-a)} \right) = +\infty$$

and

$$\lim_{k \to a} \left(\frac{(a-b)(b+2k-a)}{2(b+k-a)} \right) = \frac{a^2-b^2}{2b}$$

thus the continuous function $j(k)$ is bounded below by $\frac{a^2-b^2}{2b}$, as stated (see also Fig. 4.5). The following formula for k comes directly from (4.3) after some simple calculations, and then for h and m we can use (4.5):

$$k = \frac{j^2 - (b+j-a)^2}{2(b+j-a)}, \quad h = \frac{k(b+j)}{b-a+k+j} \text{and} \quad m = \frac{j(a-k)}{b-a+k+j}. \tag{4.8}$$

Case 4—Given a, b **and** m, **where** $0 < m < b < a$.

We proceed similarly as in Case 2. We just need to find the positive solution x to the system formed by the CL (4.2) and $y = m$. After substituting we get

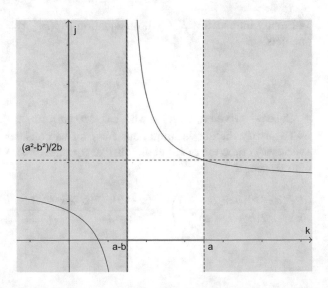

Fig. 4.5 Studying j as a function of k to justify the limitations for j

$$x^2 - x(a - b) + m^2 + m(a - b) - ab = 0,$$

and the positive solution is given by

$$h = \frac{(a - b) + \sqrt{(a + b)^2 + 4(b - m)(a + m)}}{2}.$$

Having calculated h, to find k and j formulas (4.6) and (4.7) can be used :

$$k = \frac{a^2 - m^2 - h^2}{2(a - h)}, \quad j = \frac{mk}{h - k}.$$

Case 5—Given a, k **and** h, **with** $0 < k < h < a$.

We need the positive solution y of the equation system of the circle with centre K and radius $\overline{KA}$, $(x - k)^2 + y^2 = (a - k)^2$ and the line $x = h$ (Fig. 4.6), and this yields

$$m = \sqrt{(a - k)^2 - (h - k)^2}.$$

Again we can then use

Fig. 4.6 Solving Case 5

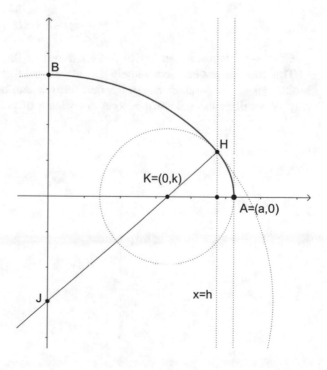

$$j = \frac{mk}{h - k}$$

and derive b, for example, from (4.3)

$$b = a - j - k + \sqrt{j^2 + k^2}. \tag{4.9}$$

Case 6—Given a, k and j, with $0 < k < a$ and $j > 0$.
Formulas already derived are enough: we need (4.9) and (4.5):

$$b = a - j - k + \sqrt{j^2 + k^2}, \quad h = \frac{k(b + j)}{b - a + k + j} \quad \text{and} \quad m = \frac{j(a - k)}{b - a + k + j}.$$

Case 7—Given a, k and m, with $0 < k < a$ and $0 < m < a - k$.
Point H has to lie on the circle with centre K and radius $\overline{KA}$; that is why it must be $m < a - k$. Intersecting this circle with the line $y = m$ yields the solution

$$h = k + \sqrt{(a - k)^2 - m^2},$$

then again formulas (4.7) and (4.9)

$$j = \frac{mk}{h - k} \quad \text{and} \quad b = a - j - k + \sqrt{j^2 + k^2}.$$

Case 8—Given a, h and j, with $0 < h < a$ and $j > 0$.
This case has not been completely solved yet, and there may also be additional limitations on the parameters. An implicit formula can be found in the following way. We call m the unknown second coordinate of point H (see Fig. 4.7) and

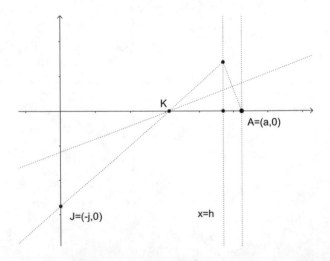

Fig. 4.7 The unsolved Case 8

calculate the x-coordinate of K by writing the equation of the axis of the segment AH

$$y - \frac{m}{2} = \frac{a - h}{m}\left(x - \frac{a + h}{2}\right)$$

and letting $y = 0$. We get for K the coordinates

$$K = \left(\frac{a^2 - h^2 - m^2}{2(a - h)}, 0\right),$$

and when we use the condition for K, J and H to be on the same line

$$\frac{y_H - y_J}{x_H - x_J} = \frac{y_K - y_J}{x_K - x_J}$$

we end up with the equation for m

$$m^3 + jm^2 + (h^2 - a^2)m - j(h - a)^2 = 0,$$

which may in general have up to three solutions such that $0 < m < a$. At this time no closed form for the solutions has been found.

   Case 9—Given a, h and m, with $0 < a - h < m < \sqrt{a^2 - h^2}$.
   Repeating Construction 9, we see that the x-coordinate of the intersection K of the axis of AH with the x-axis, which as in Case 8 is given by

$$K = \left(\frac{a^2 - h^2 - m^2}{2(a - h)}, 0\right),$$

has to be such that

$$0 < \frac{a^2 - h^2 - m^2}{2(a - h)} < h.$$

This condition becomes

$$(a - h)^2 < m^2 < a^2 - h^2,$$

which yields, since $a > h$, $0 < a - h < m < \sqrt{a^2 - h^2}$, the condition stated for m. Using again (4.7) and (4.9) we then sum up the formulas for this case:

$$k = \frac{a^2 - h^2 - m^2}{2(a - h)}, \quad j = \frac{mk}{h - k} \quad \text{and} \quad b = a - j - k + \sqrt{j^2 + k^2}. \qquad (4.10)$$

   Case 10—Given a, j and m, with $0 < m < a$ and $j > 0$.

Formulas for this case are derived using equations previously calculated. We take the second in (4.5)

$$m = \frac{j(a - k)}{b - a + k + j}$$

obtaining the following equation for k:

$$k^2(m^2 - j^2) + 2akj + j^2(m^2 - a^2) = 0$$

whose solutions are

$$k_{1;2} = \frac{-aj^2 \pm jm\sqrt{j^2 + a^2 - m^2}}{m^2 - j^2};$$

multiplication of the numerator and of the denominator by $aj + jm\sqrt{j^2 + a^2 - m^2}$ delivers the formulas

$$k_1 = \frac{j(a^2 - m^2)}{aj + m\sqrt{j^2 + a^2 - m^2}} \quad \text{and} \quad k_2 = \frac{j(a^2 - m^2)}{aj - m\sqrt{j^2 + a^2 - m^2}};$$

keeping in mind that $m < a$ it is easy to prove that $0 < k_1 < a$, therefore that it is feasible. On the other hand k_2 is negative when $m < j$, non existent when $m = j$ and greater than a if $m > j$. The formulas for this case are thus

$$k = \frac{j(a^2 - m^2)}{aj + m\sqrt{j^2 + a^2 - m^2}}, \quad j = \frac{mk}{h - k} \quad \text{and} \quad b = a - j - k + \sqrt{j^2 + k^2}.$$

Case 11—Given b, k and h, with $b > 0$ and $0 < k < h \le \sqrt{b^2 + k^2}$.

This case is the one that proved that the author's conjecture on the unicity of ovals whenever three out of six fundamental parameters are given (see beginning of Sect. 3.1) is false. We will show that certain choices of parameters b, k and h (an infinite number actually) yield two quarter ovals. Again we derive an equation for j and then observe that a real solution can be obtained under the conditions listed. Deriving a from formula (4.3) and substituting in the first of the (4.5) yields

$$h = k + \frac{k\left(b + j - \sqrt{k^2 + j^2}\right)}{\sqrt{k^2 + j^2}}$$

and eventually

$$h\sqrt{k^2 + j^2} = k(b + j).$$

Taking the square of both positive quantities we end up with the following equation for j:

$$j^2 (h^2 - k^2) - 2bjk^2 + k^2 (h^2 - b^2) = 0.$$

The two solutions are

$$j_1 = \frac{k\left(bk + h\sqrt{b^2 + k^2 - h^2}\right)}{h^2 - k^2} \quad \text{and} \quad j_2 = \frac{k\left(bk - h\sqrt{b^2 + k^2 - h^2}\right)}{h^2 - k^2}. \qquad (4.11)$$

The quantity $b^2 + k^2 - h^2$ having to be non negative, $h \leq \sqrt{b^2 + k^2}$ has to hold. The solution j_1 is clearly positive, while j_2 is positive if

$$bk - h\sqrt{b^2 + k^2 - h^2} > 0,$$

that is if $(h^2 - k^2)(h^2 - b^2) > 0$, which in our case is true if and only if $h > b$, which is of course possible. The surprising situation is therefore illustrated in Fig. 4.8. Obviously, if $b < k$ in the first place, then any feasible value for h yields two ovals. As an example of this situation let us consider the parameter values $k = 8$, $b = 10$ and $h = 12$. Formulas (4.11) give us the values

$$j_1 = \frac{4\left(10 + 3\sqrt{5}\right)}{5} \quad \text{and} \quad j_2 = \frac{4\left(10 - 3\sqrt{5}\right)}{5},$$

both positive. This means that there are two ovals with $k = 8$, $b = 10$ and $h = 12$, and they are shown in Fig. 4.9.

Case 12—Given b, k and j, **with** $b, j > 0$ **and** $0 < k < \sqrt{b^2 + 2bj}$.

The formula for a is simply derived from (4.3), and at that point the (4.5) can be used:

$$a = b + j + k - \sqrt{j^2 + k^2}, \quad h = \frac{k(b+j)}{b - a + k + j}, \quad m = \frac{j(a-k)}{b - a + k + j}. \qquad (4.12)$$

The limitation for k is explained as follows. Since k has to be such that $a - b < k < a$, then the above formula for a yields

$$j - \sqrt{j^2 + k^2} < 0 < b + j - \sqrt{j^2 + k^2}$$

which corresponds to the system

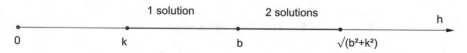

Fig. 4.8 Number of solutions to Case 11 according to the values of h

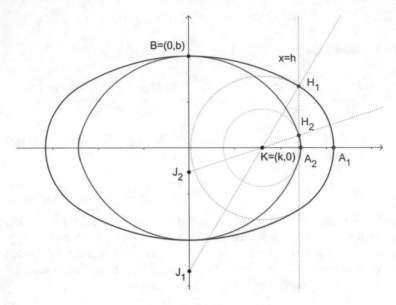

Fig. 4.9 The two ovals of Case 11 resulting from the parameter choice $k = 8$, $b = 10$ and $h = 12$.

$$\begin{cases} j - \sqrt{j^2 + k^2} < 0 \\ b + j > \sqrt{j^2 + k^2} \end{cases}$$

whose solutions are $k < \sqrt{b^2 + 2bj}$.

Case 13—Given b, k and m, with $k > 0$ and $0 < m < b$.

This case has not been completely solved yet, and there may also be additional limitations on the parameters. An implicit formula can be found in the following way. Keeping in mind Fig. 4.1 we find a general formula for h which we will use in a moment. The circle with centre J and radius $\overline{JB}$ has equation

$$x^2 + (y + j)^2 = (b + j)^2;$$

since $H = (h, m)$ has to lie on this circle, h and m must solve the equation

$$h^2 + (m + j)^2 = (b + j)^2; \tag{4.13}$$

solving for h we get

$$h = \sqrt{(b + j)^2 - (m + j)^2}. \tag{4.14}$$

We now use (4.7) to derive the following formula for k

$$k = \frac{jh}{m+j},$$ (4.15)

and substitute (4.14) in it:

$$k = \frac{j\sqrt{(b+j)^2 - (m+j)^2}}{m+j}.$$

Straightforward calculations yield for j the equation

$$2j^3(b-m) + j^2(b^2 - m^2 - k^2) - 2mjk^2 - m^2k^2 = 0,$$

which may in general have up to three positive solutions. At this moment no explicit form for the solutions has been found.

Case 14—Given b, h **and** j, **with** $b, j > 0$ **and** $0 < h < \sqrt{b^2 + bj}$. Solving (4.13) for m we get

$$m = \sqrt{(b+j)^2 - h^2} - j.$$ (4.16)

We can now use formula (4.15) and solve (4.3) for a:

$$k = \frac{jh}{m+j}, \quad a = b + j + k - \sqrt{j^2 + k^2}.$$ (4.17)

Since though $0 < m < b$ has to hold, we get

$$0 < \sqrt{(b+j)^2 - h^2} - j < b;$$

after adding j, and taking squares of the three positive members, we get

$$j^2 < b^2 + j^2 + 2bj - h^2 < b^2 + 2bj + j^2;$$

now addition of $h^2 - j^2$ yields

$$h^2 < b^2 + 2bj < b^2 + 2bj + h^2,$$

the first inequality yielding the above limitations for h, the second one being true for any value of h.

Case 15—Given b, h **and** m, **with** $h > 0$ **and** $0 < m < b < \sqrt{h^2 + m^2}$. Formula (4.16) yields

$$j + m = \sqrt{(b+j)^2 - h^2},$$

and taking squares on both sides we get eventually

$$j = \frac{m^2 + h^2 - b^2}{2(b - m)}$$

and formulas (4.17) complete the set

$$k = \frac{jh}{m + j}, \quad a = b + j + k - \sqrt{j^2 + k^2}.$$

In order for j to be positive, we have to have $m^2 + h^2 - b^2 > 0$ which yields the above condition on b.

Case 16—Given b, j and m, with $j > 0$ and $0 < m < b$.
Formulas (4.14), (4.15) and the first of the (4.12) are what we need here:

$$h = \sqrt{(b + j)^2 - (m + j)^2}, \quad k = \frac{jh}{m + j} \text{and } a = b + j + k - \sqrt{j^2 + k^2}.$$

Case 17—Given k, h and j, with $j > 0$ and $0 < k < h$.
We derive m from (4.7)

$$m = \frac{j(h - k)}{k},$$

and a from the first of (4.10):

$$k = \frac{a^2 - h^2 - m^2}{2(a - h)} \rightarrow a_{1;2} = k \pm \sqrt{k^2 + h^2 + m^2 - 2hk}$$

where only a_1 is such that $a_1 > h > k > 0$, so the acceptable value is

$$a = k + \sqrt{(h - k)^2 + m^2}; \tag{4.18}$$

finally one can use formula (4.9)

$$b = a - j - k + \sqrt{j^2 + k^2}. \tag{4.19}$$

Case 18—Given k, h and m, with $m > 0$ and $0 < k < h$.
We need (4.7), (4.18) and (4.19):

$$j = \frac{mk}{h - k}, \quad a = k + \sqrt{(h - k)^2 + m^2} \text{ and } b = a - j - k + \sqrt{j^2 + k^2}.$$

Case 19—Given k, j and m, with $k, j, m > 0$.
We invert (4.15) and use (4.18) and (4.19):

$$h = \frac{k(m+j)}{j}, \quad a = k + \sqrt{(h-k)^2 + m^2} \text{ and } b = a - j - k + \sqrt{j^2 + k^2}.$$

Case 20—Given h, j and m, with $h, j, m > 0$.
We use the first of the (4.17), and formulas (4.18) and (4.19):

$$k = \frac{jh}{m+j}, \quad a = k + \sqrt{(h-k)^2 + m^2} \text{ and } b = a - j - k + \sqrt{j^2 + k^2}.$$

We now proceed with the analysis of the three more cases we have dealt with in Chap. 3. Consistent with the numbering, we have called them Cases 23, 72 and 78.

Case 23—Given a, b and β, with $a > b > 0$ and $2\arctan\frac{a}{b} - \frac{\pi}{2} < \beta < \frac{\pi}{2}$.

To find the formulas for the remaining parameters we use Construction 23b in Chap. 3 and the corresponding Fig. 3.17. Let ϕ be the angle that BD forms with the x-axis. The equation of the line BD is

$$y - b = x \cdot \tan\phi.$$

To find point H we need to solve the system with the equation of the CL (4.2)

$$\begin{cases} y - b = x \cdot \tan\phi \\ x^2 + y^2 + x(b-a) + y(a-b) - ab = 0 \end{cases},$$

the solutions are $x_1 = 0$, the x-coordinate of B, and $x_2 = \frac{a-b-a\cdot\tan\phi-b\cdot\tan\phi}{1+\tan^2\phi}$, the x-coordinate of H. After having calculated

$$\phi = A\widehat{O}B + O\widehat{B}D = \frac{\pi}{2} + \left(\frac{1}{2}\left(\pi - \left(\frac{\pi}{2} - \beta\right)\right)\right) = \frac{3}{4}\pi + \frac{\beta}{2},$$

we get for h the following formula

$$h = \frac{a - b + (a+b) \cdot \tan\left(\frac{\pi}{4} - \frac{\beta}{2}\right)}{1 + \tan^2\left(\frac{\pi}{4} - \frac{\beta}{2}\right)}.$$

In order for BD to meet the CL inside arc AS we have to have $\frac{\pi}{2} > O\widehat{B}D > O\widehat{B}A$ which means that

$$\frac{\pi}{2} > \frac{\pi}{4} + \frac{\beta}{2} > \arctan\frac{a}{b}$$

must hold, that is for β the condition

$$2\arctan\frac{a}{b} - \frac{\pi}{2} < \beta < \frac{\pi}{2}.$$

The missing formulas for m, k and h are then copied from Case 2:

$$m = \frac{(b-a) + \sqrt{(a+b)^2 - 4h(h-a+b)}}{2},$$

$$k = \frac{a^2 - m^2 - h^2}{a(a-h)} \quad \text{and} \quad j = \frac{mk}{h-k}.$$

Case 72—Given k, j and p, with k, $j > 0$ and

$$\frac{\sqrt{j^2 + k^2} - j}{k} < p < 1$$

To find a formula for a we just consider that $b = ap$ and substitute it inside (4.3). What we get is

$$a = \frac{k + j - \sqrt{j^2 + k^2}}{1 - p},$$

and for the remaining parameters use from Case 6

$$b = a - j - k + \sqrt{j^2 + k^2}, \quad h = \frac{k(b+j)}{b-a+k+j} \quad \text{and} \quad m = \frac{j(a-k)}{b-a+k+j}.$$

In order to explain the limitation for p we can repeat Construction 72. We need to find the centre C of the CL. Let $\alpha = \arctan\frac{j}{k}$ (see from now on Fig. 4.10). The equation of the bisector of α is $y = \tan\frac{\alpha}{2}(x - k)$; we find C by intersection with $y = -x$ (see Theorem 2.3):

$$C = \left(\frac{k \cdot \tan\frac{\alpha}{2}}{1 + \tan\frac{\alpha}{2}}, -\frac{k \cdot \tan\frac{\alpha}{2}}{1 + \tan\frac{\alpha}{2}} \right).$$

We now need the line through C having slope $m = 1$ and we need to intersect it with the line r through O with given slope p in order to find T: the solution to the system

$$\begin{cases} y = px \\ y + \dfrac{k \cdot \tan\dfrac{\alpha}{2}}{1 + \tan\dfrac{\alpha}{2}} = x - \dfrac{k \cdot \tan\dfrac{\alpha}{2}}{1 + \tan\dfrac{\alpha}{2}} \end{cases} \tag{4.20}$$

exists and it is feasible if r has a slope p lesser than 1, in order to intersect the line coming from C, and on the other hand greater than the slope of the line OG (see

Fig. 4.10 Finding the
limitations for p in Case 72

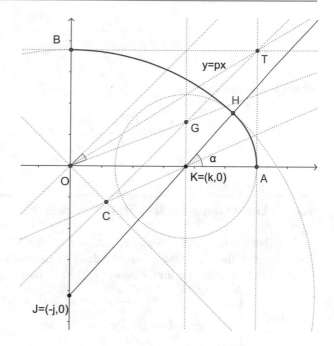

Fig. 4.10), in order for T to have an x-coordinate greater than k. The coordinates of
G are, substituting k into the second equation in (4.20):

$$G = \left(k, k\left(\frac{1 - \tan\frac{\alpha}{2}}{1 + \tan\frac{\alpha}{2}} \right) \right),$$

that is OG has slope

$$m_{OG} = \left(\frac{1 - \tan\frac{\alpha}{2}}{1 + \tan\frac{\alpha}{2}} \right). \tag{4.21}$$

We now need to express $\tan\frac{\alpha}{2}$ in terms of j and k. Since $\tan\alpha = \frac{j}{k}$ we get, from well-known trigonometry formulas,

$$\sin\alpha = \frac{\tan\alpha}{\sqrt{1 + \tan^2\alpha}} = \frac{j}{\sqrt{j^2 + k^2}} \quad \text{and} \quad \cos\alpha = \frac{1}{\sqrt{1 + \tan^2\alpha}} = \frac{k}{\sqrt{j^2 + k^2}},$$

and consequently that

$$\tan\frac{\alpha}{2} = \frac{1 - \cos\alpha}{\sin\alpha} = \frac{\sqrt{j^2 + k^2} - k}{j}. \tag{4.22}$$

Setting now $m_{OG} < p < 1$, and using (4.21) and (4.22) one gets

$$\frac{1 - \frac{\sqrt{j^2 + k^2} - k}{j}}{1 + \frac{\sqrt{j^2 + k^2} - k}{j}} < p < 1.$$

which leads after some calculation to

$$\frac{\sqrt{j^2 + k^2} - j}{k} < p < 1.$$

4.2　Limitations for the Frame Problem

Let's go back to the Direct Frame Problem of Sect. 3.3. We want to find proof of the limitations for the point which should belong to an oval with given axes.

Case 117—Given a, b **and a point** $(\bar{x}, \bar{y})$ **such that**

either a) $\begin{cases} \bar{x}^2 + \bar{y}^2 + \bar{y}\left(\dfrac{a^2 - b^2}{b}\right) - a^2 > 0 \\ \bar{x}^2 + \bar{y}^2 + \bar{x}(b - a) + \bar{y}(a - b) - ab < 0 \\ 0 < \bar{y} < b \end{cases}$

or b) $\begin{cases} \bar{x}^2 + \bar{y}^2 + 2\bar{x}(b - a) + a^2 - 2ab < 0 \\ \bar{x}^2 + \bar{y}^2 + \bar{x}(b - a) + \bar{y}(a - b) - ab > 0 \end{cases}.$

We are looking for an oval with half axes a and b through a given point. Such a point will have to belong to either an arc with a J centre or to an arc with a K centre (as a special case to both).

Case 117a We want to show that for the oval to be drawn and for $M = (\bar{x}, \bar{y})$ to belong to an arc with a longer radius, conditions a) have to be met; in such a case Construction 117a in Chap. 3 works.

The equation of the axis of segment BM (see Fig. 4.11) is

$$y - \frac{\bar{y} + b}{2} = \frac{\bar{x}}{b - \bar{y}}\left(x - \frac{\bar{x}}{2}\right)$$

and its intersection with the y-axis is

$$\left(0, \frac{\bar{y} + b}{2} + \frac{\bar{x}^2}{2(b - \bar{y})}\right).$$

For this point to be a feasible J point, the condition of Case 3 has to hold:

$$\frac{\bar{y} + b}{2} + \frac{\bar{x}^2}{2(b - \bar{y})} < -\frac{a^2 - b^2}{2b},$$

and after some simple calculations this becomes

Fig. 4.11 Finding the limitations for $M = (\bar{x}, \bar{y})$ in Case 117a

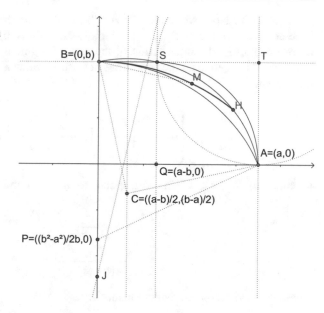

$$\bar{x}^2 + \bar{y}^2 + \bar{y}\left(\frac{a^2 - b^2}{b}\right) - a^2 > 0,$$

which is the set of points outside the circle with centre

$$P = \left(\frac{b^2 - a^2}{2b}, 0\right)$$

and radius

$$\frac{b^2 + a^2}{2b} = \overline{PB},$$

P being the point on BO such that $\overline{PB} = \overline{BA}$.

On the other hand the CL and the circle with centre J and radius $\overline{JB}$ both run through H and B, so the two AS arcs are one outside the other. Since S is a point on the CL which can easily be proved to be outside the circle with centre J and radius $\overline{JB}$, then the same has to hold for any points on the two arcs. This means that $M(\bar{x}; \bar{y})$ is *inside* the CL, and this condition, remembering (4.2), can be written

$$\bar{x}^2 + \bar{y}^2 + \bar{x}(b - a) + \bar{y}(a - b) - ab < 0.$$

Case 117a corresponds thus to choosing $M = (\bar{x}, \bar{y})$ inside the blue area in Fig. 3.37.

Case 117b. Now we show that for the oval to be drawn and for $M = (\bar{x}, \bar{y})$ to belong to an arc with smaller radius, conditions b) have to be met. The corresponding K, the intersection of the axis of segment AM with the horizontal axis, would have to meet the conditions of Case 1, which mean, for the solution x of the system

$$\begin{cases} \dfrac{y - \dfrac{\bar{y}}{2} = a - \bar{x}}{\bar{y}} \\ \left(x - \dfrac{a + \bar{x}}{2} \right) y = 0 \end{cases}$$

to satisfy $a - b < x < a$. This corresponds to finding $\bar{x}$ and $\bar{y}$ such that

$$\begin{cases} a - b < \dfrac{a^2 - \bar{x}^2 - \bar{y}^2}{2(a - \bar{x})} \\ \dfrac{a^2 - \bar{x}^2 - \bar{y}^2}{2(a - \bar{x})} < a \end{cases} ;$$

calculations yield the equivalent inequality

$$\bar{x}^2 + \bar{y}^2 + 2\bar{x}(b - a) + a^2 - 2ab < 0,$$

which is half of what we wanted to prove. Now just note that the CL and the circle having centre any feasible $K = (k, 0)$ and radius $a - k$, always meet only in A and H, as seen in Chap. 2. These two points divide the circle with centre K into two arcs, one of which is inside the CL and the other outside. Being, for example, P in Fig. 2.9 inside the CL, since it belongs to the chord BH, and because the connection point H is between M and P, then $M = (\bar{x}, \bar{y})$ has to be outside, which means that

$$\bar{x}^2 + \bar{y}^2 + \bar{x}(b - a) + \bar{y}(a - b) - ab > 0$$

has to hold.

Case 118a. Finding an oval circumscribing a rectangle with arcs with smaller radius through the vertices. To prove the limitations for a in the first part of the Inverse Frame Problem of Chap. 3 we take a generic point $P = (\bar{x}, \bar{y})$, representing the vertex of the inscribed rectangle (Fig. 4.12, top). A feasible $A = (a, 0)$ is a point such that $a > \bar{x}$ and such that the axis of segment PA lands on the horizontal axis—to produce point K—between the centre of symmetry and the right edge of the rectangle. The equation of the axis is

$$\dfrac{y - \dfrac{\bar{y}}{2} = a - \bar{x}}{\bar{y}\left(x - \frac{a + \bar{x}}{2} \right)},$$

and the point where it intersects the horizontal axis has the coordinate

Fig. 4.12 Finding the
limitations for a and then for
b in Case 118a (top); finding
the limitations for b and then
for a in Case 118b (bottom)

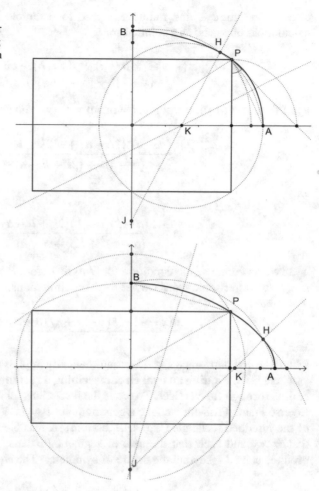

$$k = \frac{\bar{y}^2 + \bar{x}^2 - a^2}{2(\bar{x} - a)};$$

imposing for this value to be between 0 and $\bar{x}$ yields for a the cited limitations

$$\sqrt{\bar{x}^2 + \bar{y}^2} < a < \bar{x}.$$

To find the resulting limitations for b we must first derive a general formula for
h as a function of a, b and k. To do so we look for the intersection of the CL with the

circle with centre K and radius $\overline{KA}$ (see for example Fig. 4.4). After lengthy calculations one gets, as a solution for

$$\begin{cases} x^2 + y^2 + x(b-a) + y(a-b) - ab = 0 \\ (x-k)^2 + y^2 = (a-k)^2 \end{cases},$$

apart from $A = (a, 0)$, the point $H = (h, m)$ with coordinates

$$h = \frac{(a-b)^2(2k-a) + a(2k+b-a)^2}{(a-b)^2 + (2k+b-a)^2}$$

and

$$m = \frac{2(a-k)(a-b)(2k+b-a)}{(a-b)^2 + (2k+b-a)^2}.$$

If we now impose the condition for H to lie before P but beyond K, that is that $k < h < \bar{x}$, after some work we get for b the limitations

$$a - k < b < \frac{a(\bar{x}-k) - k\left(\bar{x}-a+\sqrt{(a-\bar{x})(a+\bar{x}-2k)}\right)}{\bar{x}-k},$$

where the first part, since $\bar{y} < a - k$ automatically implies that $b > \bar{y}$ as it should be.

Case 118b. Finding an oval circumscribing a rectangle with arcs with longer radius through the vertices. To prove the limitations for b in the second part of the Inverse Frame Problem we take a generic point $P = (\bar{x}, \bar{y})$, representing the vertex of the inscribed rectangle (Fig. 4.12, bottom). A feasible $B = (0, b)$ is a point such that $b > \bar{y}$ and such that the axis of segment PB lands on the vertical axis—to produce point J—beyond the centre of symmetry. The equation of such axis is

$$y - \frac{\bar{y}+b}{2} = \frac{\bar{x}}{b-\bar{y}}\left(x - \frac{\bar{x}}{2}\right)$$

and it intersects the vertical axis in

$$j = \frac{\bar{x}^2 - b^2 + \bar{y}^2}{2\bar{x}};$$

imposing $j > 0$ yields for b the condition

$$\bar{y} < b < \sqrt{\bar{x}^2 + \bar{y}^2},$$

represented as a green segment in Fig. 4.12 (bottom).

To find the resulting limitations for a this time we look for the intersection of the CL with the circle with centre J and radius $\overline{JB}$. The other solution of

$$\begin{cases} x^2 + y^2 + x(b-a) + y(a-b) - ab = 0 \\ x^2 + (y+j)^2 = (b+j)^2 \end{cases},$$

apart from $B = (0, b)$, is found as the point $H = (h, m)$ with coordinates

$$h = \frac{(a-b)(b+j)(b-a+2j)}{(a-b-j)^2 + j^2} \quad \text{and} \quad m = \frac{j(2bj + b^2 - a^2)}{(a-b-j)^2 + j^2}.$$

If we now impose the condition for H to lie beyond P but to remain inside the first quadrant, that is that $0 < m < \bar{y}$, after some work we get for a the limitations

$$\frac{\bar{y}(j+b) + \sqrt{(b-\bar{y})(b+\bar{y}+2j)}}{j+\bar{y}} < a < \sqrt{b(2j+b)},$$

also represented as a green segment in Fig. 4.12 (bottom).

4.3   Measuring a Four-Centre Oval

We obtain here formulas for the length of the perimeter and for the area of an oval.

The length of an arc of a circle with radius r and angle at the centre α is given by $l = r\alpha$, if α is measured in radiants (otherwise multiply by $\frac{\pi}{180°}$). So in Fig. 4.13 we can calculate arcs $l_1 = AH$ and $l_2 = BH$ as follows:

Fig. 4.13 Calculating perimeter and area of a simple oval

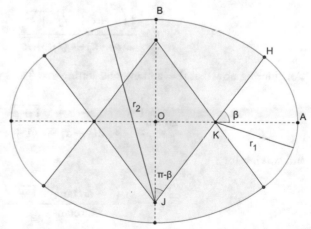

$$l_1 = \beta \cdot r_1 \text{ and } l_2 = \left(\frac{\pi}{2} - \beta\right) \cdot r_2.$$

So a straightforward formula for the whole length is

$$L = 2\pi r_2 - 4\beta(r_2 - r_1).$$

In order to get formulas for the whole length depending on three of the fundamental parameters we remember that

$$r_1 = a - k, \quad r_2 = a - k + \sqrt{j^2 + k^2} \text{ and } \beta = \arctan\frac{j}{k},$$

using (4.1) and Fig. 4.13. This yields for the whole oval the formula

$$L = 4(l_1 + l_2) = 4\left[\frac{\pi}{2}(a - k) + \left(\frac{\pi}{2} - \arctan\frac{j}{k}\right)\sqrt{j^2 + k^2}\right]$$

An interesting problem is that of finding an oval where point H divides the quarter oval into two arcs of equal length. This is obviously achieved when

$$\beta \cdot r_1 = \left(\frac{\pi}{2} - \beta\right) \cdot r_2.$$

Using the same formulas as above for r_1 and r_2, we get

$$\beta\left(2a - 2k + \sqrt{j^2 + k^2}\right) = \frac{\pi}{2}\left(a - k + \sqrt{j^2 + k^2}\right),$$

and so for β

$$\beta = \frac{\frac{\pi}{2}\left(a - k + \sqrt{j^2 + k^2}\right)}{2(a - k) + \sqrt{j^2 + k^2}}.$$

Considering now that $\beta = \arctan\frac{j}{k}$, and letting $z = \frac{k}{j}$ and $s = \frac{a}{j}$ we obtain

$$\arctan\frac{1}{z} = \frac{\pi}{2} \cdot \frac{s - z + \sqrt{1 + z^2}}{2(s - z) + \sqrt{1 + z^2}},$$

and solving for s

$$s = z + \frac{\left(\frac{\pi}{2} - \arctan\frac{1}{z}\right)\sqrt{1 + z^2}}{2\arctan\frac{1}{z} - \frac{\pi}{2}}.$$

Not all values for z yield a positive value for s, but those in the interval $\left[\frac{\sqrt{3}}{3}; 1\right[$ definitely do, since the above formula can be written

$$s = \frac{\arctan\frac{1}{z}\left(2z - \sqrt{1+z^2}\right) + \frac{\pi}{2}\left(\sqrt{1+z^2} - z\right)}{2\arctan\frac{1}{z} - \frac{\pi}{2},}$$

and the first term in the nominator is not negative for $z \geq \frac{\sqrt{3}}{3}$, the second term is always strictly positive and the denominator is positive for $z < 1$.

So, in order to find such an oval *shape* one chooses a value for z and calculates s. For $z = \frac{\sqrt{3}}{3}$ for example, corresponding to an angle $\beta = \frac{\pi}{3}$, one gets $s = \sqrt{3}$. This means that, given a value for a, the oval having $j = \frac{a}{s} = \frac{\sqrt{3}}{3}a$ and $k = jz = \frac{a}{3}$ is an oval where the connection point H divides the quarter oval into two arcs of equal length. This is the famous 4th construction by Serlio, which is described in Chap. 6 among other remarkable four-centre ovals. It is this property that made Borromini choose it as a guideline for his plan of the dome of the church of *San Carlo alle Quattro Fontane* in Rome, as suggested in Chap. 7. But this is not the only possible choice: a choice of $z = \frac{3}{4}$ yields approximate values, given a, of $j \approx 0.28 \cdot a$ and $k \approx 0.21 \cdot a$, forming another oval shape where the connection point H divides the quarter oval into two arcs of equal length.

For the area of an oval we use the formula for a sector, which is $S = \frac{\alpha \cdot r^2}{2}$. From Fig. 4.13 we get, for the sectors corresponding to arcs AH and BH,

$$S_1 = \frac{(a-k)^2}{2}\arctan\frac{j}{k} \quad \text{and} \quad S_2 = \frac{(b+j)^2}{2}\left(\frac{\pi}{2} - \arctan\frac{j}{k}\right).$$

Adding the two sectors, subtracting the area of the triangle OJK and then multiplying by 4 we get the area of the whole oval

$$S = (b+j)^2\left(\pi - 2\arctan\frac{j}{k}\right) + 2(a-k)^2\arctan\frac{j}{k} - 2jk.$$

To obtain a formula with only three parameters we use (4.9) for b, and get

$$S = \left(a - k + \sqrt{j^2 + k^2}\right)^2\left(\pi - 2\arctan\frac{j}{k}\right) + 2(a-k)^2\arctan\frac{j}{k} - 2jk.$$

The version depending on parameters r_1, r_2 and β is also easily obtained:

$$S = \frac{\pi \cdot r_2{}^2}{4} - \frac{\beta \cdot (r_2{}^2 - r_1{}^2)}{2} - \frac{(r_2 - r_1)^2 \sin\beta \cdot \cos\beta}{2}$$

4.4   Concentric Ovals

Given an oval with parameters a', b', k', h', j' and m', a whole set of other ovals can be drawn just by keeping fixed k' and j' and adding to a' and b' the same amount r, which will turn the starting oval into another one with four arcs made of the two starting radii both increased by r. These are called *concentric ovals* in the sense that they share the same centres J and K for the arcs that form them, but as we will see, other than concentric circles, they do not have the same shape and their axes change their ratio all the time.

There is a limit for ovals lying *inside* the given one since k cannot exceed a, while concentric ovals outside the first one can be as big as one wishes.

In Fig. 4.14 we consider the starting oval with its symmetry centre on the origin of a coordinate system whose axes coincide with the oval axes. The main points forming the quarter oval have coordinates

$$A' = (a',0), \quad B' = (0,b'), \quad K' = (k',0), \quad J' = (0,-j') \text{ and } H' = (h',m').$$

Taking then any number r (positive or negative) such that $r > k - a$, we find a concentric oval just by adding r to a' (or b') getting

$$A = (a'+r,0), \quad B = (0,b'+r), \quad K = (k',0) \text{ and } J = (0,-j');$$

while the new connection point H has coordinates (see (4.5))

Fig. 4.14 The first of a number of concentric ovals

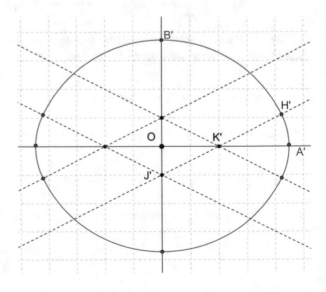

$$h = \frac{k'(b'+r+j')}{b'-a'+k'+j'} = \frac{k'(b'+j')}{b'-a'+k'+j'} + \frac{rk'}{b'-a'+k'+j'} =$$

$$= h' + \frac{rk'}{b'-a'+k'+j'}$$

and

$$m = \frac{j'(a'+r-k')}{b'-a'+k'+j'} = m' + \frac{rj'}{b'-a'+k'+j'},$$

see Fig. 4.15.

The shape of this oval is rounder, when r is positive, while the two ovals are *parallel* in the sense that the area between them is scanned by a segment of constant length r. This was the idea, for example, when planning Roman amphitheatres, although eight-centre ovals were also often used (see Sect. 8.2 on the Colosseum).

The new value of the axis ratio p is

$$p = \frac{b'+r}{a'+r},$$

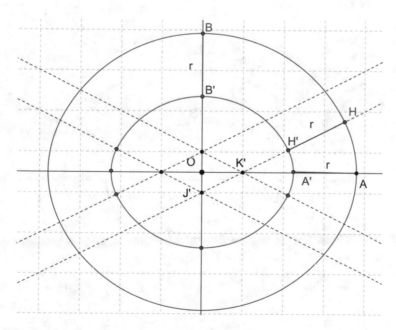

Fig. 4.15 Constructing an oval concentric with the first one

which means that, as a function of r, its graph is the one in Fig. 4.16. The function nears itself to the value 1, which means that for big values of r the oval resembles more and more a circle.

A selection of concentric ovals is displayed in Fig. 4.17.

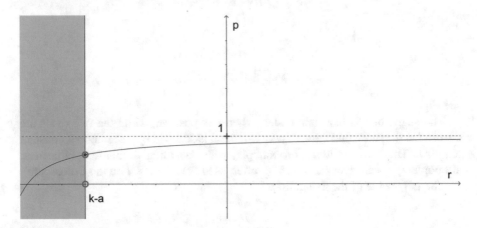

Fig. 4.16 The graph of $p(r)$ for concentric ovals

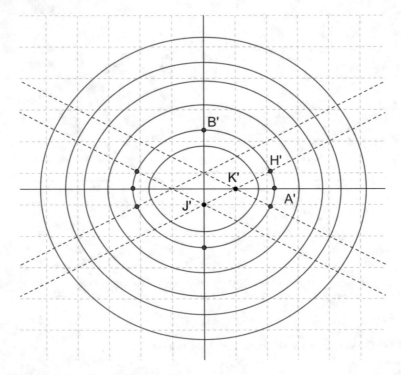

Fig. 4.17 A selection of concentric ovals

References

1. Dotto, E.: Note sulle costruzioni degli ovali a quattro centri. Vecchie e nuove costruzioni dell'ovale. Disegnare Idee Immagini. **XII**(23), 7–14 (2001)
2. López Mozo, A.: Oval for any given proportion in architecture: a layout possibly known in the sixteenth century. Nexus Netw. J. **13**(3), 569–597 (2011)
3. Mazzotti, A.A.: What Borromini might have known about ovals. Ruler and compass constructions. Nexus Netw. J. **16**(2), 389–415 (2014)
4. Ragazzo, F.: Geometria delle figure ovoidali. Disegnare idee immagini. **VI**(11), 17–24 (1995)
5. Rosin, P.L.: A family of constructions of approximate ellipses. Int. J. Shape Model. **8**(2), 193–199 (1999)

Optimisation Problems for Ovals

Once formulas describing parameters or other characteristics of an oval have been derived (in Chap. 4), calculus can be used to consider them as functions of a certain quantity and see how they vary. In this short chapter as an example we present here the solution to the problems of minimising the difference and the ratio of the radii of a simple oval with given axes a and b. The first of these two problems being suggested by Edoardo Dotto.

We know from (4.1) that $r_2 = b + j$ and $r_1 = a - k$. We will study $d = r_2 - r_1$ as a function of j. Letting $q = a - b$ we get

$$d = b + j - a + k = j + k - q,$$

then using the first of (4.8) we get

$$k = \frac{j^2 - (b + j - a)^2}{2(b + j - a)} = \frac{j^2 - (j - q)^2}{2(j - q)} = \frac{q(2j - q)}{2(j - q)}$$

and so for d, after simple calculations,

$$d = j - q + \frac{q(2j - q)}{2(j - q)} = j + \frac{q^2}{2(j - q)}.$$

In this context the function

$$d(j) = j + \frac{q^2}{2(j - q)}$$

is defined in the open interval $D =]\dfrac{a^2 - b^2}{2b}; +\infty[$ (see Case 3 in Sect. 4.1) and its graph is contained in the I quadrant. We perform the limits as j approaches the endpoints of D:

© Springer International Publishing AG 2017
A.A. Mazzotti, *All Sides to an Oval*, DOI 10.1007/978-3-319-39375-9_5

$$\lim_{j \to \frac{a^2-b^2}{2b}} d(j) = \frac{a^2 - b^2}{2b} + \frac{(a-b)^2}{2\left(\frac{a^2-b^2}{2b} - a + b\right)} = \frac{a^2 - b^2}{2b} + b = \frac{a^2 + b^2}{2b}$$

and

$$\lim_{j \to +\infty} d(j) = +\infty.$$

We now differentiate with respect to j:

$$d'(j) = 1 - \frac{q^2}{2(j-q)^2} = \frac{2j^2 - 4jq + q^2}{2(j-q)^2},$$

and find that the function decreases until $j_2 = q\left(1 + \frac{\sqrt{2}}{2}\right)$ and then increases as $j \to +\infty$ (note that the other root of the derivative $j_1 = q\left(1 - \frac{\sqrt{2}}{2}\right)$ is never feasible since $j_1 < q = a - b = \frac{(a^2 - b^2)}{(a+b)} < \frac{(a^2 - b^2)}{2b}$).

So the function $d(j)$ has a minimum for

$$\bar{j} = (a-b)\left(1 + \frac{\sqrt{2}}{2}\right)$$

only if $j \in D$, i.e. if

$$(a-b)\left(1 + \frac{\sqrt{2}}{2}\right) > \frac{(a^2 - b^2)}{2b} \tag{5.1}$$

otherwise d decreases as $j \to \frac{a^2 - b^2}{2b}$ but never reaches a minimum value because $\frac{a^2 - b^2}{2b}$ is an unfeasible value for j. Condition (5.1) yields

$$\frac{a}{b} < 1 + \sqrt{2}. \tag{5.2}$$

This means that ovals satisfying (5.2) allow for an oval with the minimum difference between the radii, the others don't. Figure 5.1 shows the two different cases. The oval with the minimum radius difference has a value d of

$$d(j_{MIN}) = d\left((a-b)\left(1 + \frac{\sqrt{2}}{2}\right)\right) = (a-b)\left(1 + \sqrt{2}\right).$$

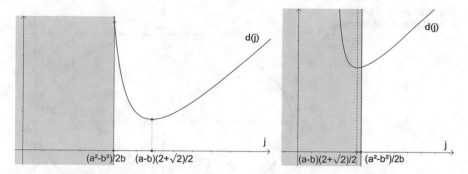

Fig. 5.1 The graph of $d(j)$ when $a = 7$ and $b = 5$ and when $a = 8$ and $b = 3$

The corresponding value of k cam be found substituting $q\left(1 + \dfrac{\sqrt{2}}{2}\right)$ for j in (4.3) and solving for k:

$$a = b + (a - b)\left(\frac{2 + \sqrt{2}}{2}\right) + k - \sqrt{(a - b)^2\left(\frac{2 + \sqrt{2}}{2}\right)^2 + k^2}$$

and what one gets is

$$\bar{k} = (a - b)\left(1 + \frac{\sqrt{2}}{2}\right) = \bar{j},$$

which means that such an oval has $\beta = \dfrac{\pi}{4}$ and when a and b satisfy (24) it is very easy to draw (Fig. 5.2), either using Constructions (23a) and (23b) (see Chap. 3), or using a method developed by Lopez Mozo (see [2]).

Finding the oval of given axes minimizing the ratio of the two radii makes more sense if one is interested in the oval which is rounder. We can proceed as follows.

Consider a and b as fixed. Using for r_1, r_2 and k the same formulas as above, we get for the ratio t of the radii, as a function of j,

$$t(j) = \frac{b + j}{a - k} = \frac{b + j}{a - \frac{j^2 - (b + j - a)^2}{2(b + j - a)}} = \frac{2(b + j)(b + j - a)}{b^2 - a^2 + 2bj};$$

this function is also defined in $D = \left]\dfrac{a^2 - b^2}{2b}; +\infty\right[$ and taking the limits to the endpoints of D we get

$$\lim_{j \to \frac{a^2 - b^2}{2b}} t(j) = \lim_{j \to +\infty} t(j) = +\infty.$$

Differentiation of $t(j)$ yields, after some calculations

Fig. 5.2 An oval with
minimum radius difference,
for feasible values of *a* and *b*

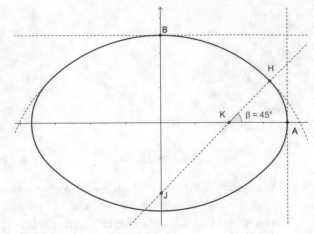

$$t'(j) = \frac{4bj^2 + 4j(b^2 - a^2) + 2a(a-b)^2}{b^2 - a^2 + 2bj};$$

considering that the denominator is always positive in D it is enough to study the sign of the numerator. Of the two roots of the numerator, the only one inside D for any values of *a* and *b* is

$$\hat{j} = \frac{(a-b)\left(a + b + \sqrt{a^2 + b^2}\right)}{2b}, \tag{5.3}$$

and the function $t(j)$ reaches its minimum there. The corresponding value of k is

$$\hat{k} = \frac{(a-b)\left(a + \sqrt{a^2 + b^2}\right)}{\left(a - b + \sqrt{a^2 + b^2}\right)}$$

while the minimum value of the ratio is

$$t\left(\hat{j}\right) = \frac{(a-b)\sqrt{a^2 + b^2} + \left(a^2 + b^2 - ab\right)}{b^2}.$$

When we now calculate $\dfrac{\hat{j}}{\hat{k}}$ using the derived formulas we get, after some calculation,

$$\frac{\hat{j}}{\hat{k}} = \frac{a}{b},$$

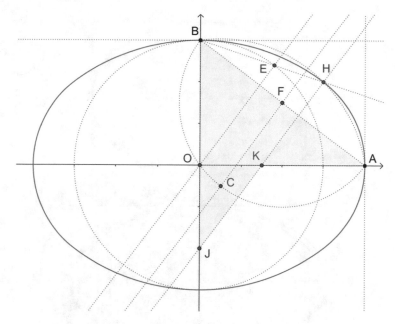

Fig. 5.3 Constructing an oval with minimum ratio between the radii

which means that for given a and b, the oval with minimum radius ratio is the only one where the lines AB and JK are orthogonal, or in other words, the oval where the triangles OAB and OJK have the same angles.

The remaining parameters can be calculated by means, for example, of formulas (4.5).

Figure 5.3 shows the very easy construction which makes use of Construction 23b in Chap. 3, and of the fact that for such an oval

$$\beta = \arctan\frac{\widehat{j}}{k} = \arctan\frac{a}{b}.$$

Construction of the oval with minimum radius ratio, for any given a and b (see the link to the Geogebra animation video on www.mazzottiangelo.eu/en/pcc. asp):

– draw segment AB and the half-circle with diameter AB containing O
– let C be the point on this half circle dividing it into two equal parts
– draw the perpendicular to AB through O and let E be its intersection with the circle with centre O and radius $\overline{OB}$
– find the intersection H between the Connection Locus—the arc of circle AB with centre C—and the line EB
– H is a connection point for the oval, and the corresponding line of the centres is the parallel to OE through H

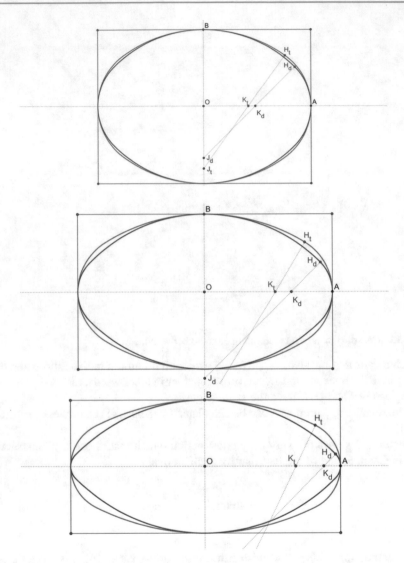

Fig. 5.4 Comparing the oval with minimum radius difference (in red, with d as subscript for the corresponding parameters) to the oval with minimum radius ratio (in green, with t as subscript for the corresponding parameters), for three different choices of a and b

The above construction means that there are an infinite number of oval shapes with OAB similar to OJK, one for every value of $p < 1$, since the above construction, corresponding to Case 23 in Chap. 3 always works because the choice of $\beta = \arctan\frac{a}{b}$ automatically verifies the condition for β.

In Chap. 6 we will present three such ovals, the Diatessaron oval, the inscribed equilateral oval, and an oval presented by the author.

Figure 5.4 compares the oval with minimum radius difference with the oval with minimum radius ratio, given the same a and b. In all three examples a and b satisfy condition (5.2).

In [1] Fernández Gómez proposes the construction of two particular ovals given any values for a and b. These two ovals are actually the minimum radius difference oval and the minimum radius ratio oval although the author does not know it. Both constructions are as straightforward as the ones presented here.

Formulas proved in Chap. 4 allow for more similar investigations.

References

1. Fernández Gómez, M.: Trazas de óvalos y elipses en los Tratados de Arquitectura de los siglos XVI y XVII. In: La formación cultural arquitectónica en la enseñanza del dibujo, Actas del Quinto Congreso Internacional de Expresión Gráfica Arquitectónica, Universidad de las Palmas de Gran Canaria, Departamento de Expresión Gráfica y Proyectación Arquitectónica. Las Palmas de Gran Canaria, 5th–7th May 1994, pp. 335–378 (1994)
2. López Mozo, A.: Oval for any given proportion in architecture: a layout possibly known in the sixteenth century. Nexus Netw. J. **13**(3), 569–597 (2011)

Remarkable Four-Centre Oval Shapes

6

There are ∞^3 simple ovals of given axis lines, and since any of them has infinite versions with the same shape (you just need to multiply all parameters except p and β by the same positive number), we can say that there are ∞^2 different shapes of four-centre ovals. What we have dealt with in Chaps. 3–6 are general purpose constructions and formulas, where you can end up with any oval varying the parameter values. This chapter is dedicated to particular oval shapes, chosen either for their use in architecture and/or for their geometric elegance. Parameter formulas in Chap. 4 help in investigating the properties of these specific oval shapes. If, on the other hand, one fixes some kind of relation between two parameters, or fixes a value for p or for β, then one gets an oval shape family of ∞^1 shapes with something in common. Cases in the literature include Serlio's first construction (see [13]), the oval shape families by Gridgeman and Franchi (used to fit a given ellipse—see [9] for references and comments), those by Bianchi, Kitao and the third study by Hewitt (see [10] for references and comments) as well as studies by Zerlenga (in [16]) of variations of single parameters and borderline cases of Bosse'e constructions. In this book both the minimal radius ratio and the minimal radius difference can be considered examples of oval shape families.

The main references for this chapter are [3, 8, 10, 13]. References [2, 5] also contain a selection of oval shapes in use in architecture the sixteenth and seventeenth centuries. The order follows no specific logic. The point and the parameter names are those used in the preceding chapters.

The celebrated work by Serlio [13] displays four different constructions (1545). Since the first one is a method for drawing a family of infinite (∞^1) oval shapes, we will start by his second construction.

Serlio 2 A circle is drawn and the four intersections with the axes are chosen as the four centres for the arcs. The arcs with centres on the longer axis are chosen on circles identical to the previous one (see Fig. 6.1). Given then a generic value for k, we use $\beta = \frac{\pi}{4}$ and $r_1 = k$ to obtain, also via the formulas in Chap. 4.

© Springer International Publishing AG 2017
A.A. Mazzotti, *All Sides to an Oval*, DOI 10.1007/978-3-319-39375-9_6

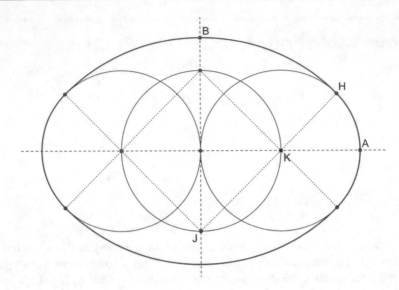

Fig. 6.1 Serlio's second construction

$$a = 2k, \ b = k\sqrt{2}, \ j = k, \ h = k\left(\frac{2+\sqrt{2}}{2}\right), \ m = k\frac{\sqrt{2}}{2}, \ r_2 = k(1+\sqrt{2}), \ p$$

$$= \frac{\sqrt{2}}{2}.$$

Since $\beta = \frac{\pi}{4}$, as shown in Chap. 5, this is the oval form minimizing the difference between the radii whenever $p = \frac{\sqrt{2}}{2}$.

Serlio 3 Two adjacent squares are drawn (Fig. 6.2). Their centres and the vertices of the common sides are chosen as centres for the arcs, while the four remaining vertices are chosen as connection points. This means that given k, we have $\beta = \frac{\pi}{4}$ and $r_1 = k\sqrt{2}$; the remaining parameters are

$$a = k(1+\sqrt{2}), \ b = k(2\sqrt{2}-1), \ j = k,$$
$$h = 2k, \ m = k, \ r_2 = 2k\sqrt{2}, \ p = 5 - 3\sqrt{2}.$$

Again, this construction minimizes the difference between the radii whenever $p = 5 - 3\sqrt{2}$.

The following is also known as the equilateral oval (see for example [3]). In Chap. 7 we will use the term canonic translating the adjective "canonico" widely used in the Italian literature on the subject of baroque architecture, and show how this shape has been the invisible reference for Borromini's dome of San Carlo alle Quattro Fontane in Rome.

Serlio 4 Once a centre of an arc with the smaller radius has been chosen (Fig. 6.3), and its symmetrical w.r.t. the vertical axis, two circles are drawn having centres in

Fig. 6.2 Serlio's third construction

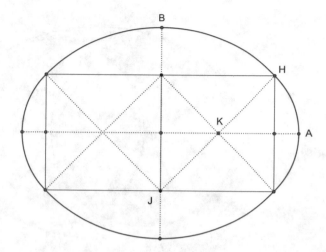

each one of these through the other one. Their intersections are chosen as centres of the other arcs. Two equilateral triangles are formed. From $\beta = \frac{\pi}{3}$ and $r_1 = 2k$ we get

$$a = 3k, \quad j = k\sqrt{3}, \quad h = 2k, \quad m = k\sqrt{3},$$

$$b = k(4 - \sqrt{3}), \quad r_2 = 4k, \quad p = \frac{4 - \sqrt{3}}{3}.$$

Point H divides the top right-hand side quarter oval into two arcs of equal length (see Sect. 4.3).

Extensions of Serlio's ideas concerning inner circles and squares have been presented by Rosin in [10]. We display them here as Rosin 1 and Rosin 2.

Rosin 1 This time three squares adjacent to one another are drawn. The two arcs with the smaller radius have centres at the centre of the left and right squares, while the other centres are on the lines of their diagonals. The connection points are chosen on the other vertices of these diagonals (Fig. 6.4). From $\beta = \frac{\pi}{4}$ and $r_1 = \frac{k\sqrt{2}}{2}$ we get

$$a = \frac{k(2 + \sqrt{2})}{2}, \quad j = k, \quad h = \frac{3}{2}k, \quad m = \frac{k}{2},$$

$$b = \frac{k(3\sqrt{2} - 2)}{2}, \quad r_2 = \frac{3k\sqrt{2}}{2}, \quad p = 4\sqrt{2} - 5$$

And this is another oval enjoying the minimal radius difference property.

Rosin 2 Two more circles are added to Serlio's fourth construction, and these are used for two of the arcs. The centres of the other two arcs are chosen as in Serlio's version (Fig. 6.5). Given k we have $a = \frac{5}{3}k$ and $j = \frac{k\sqrt{3}}{3}$, from which we calculate

$$r_1 = \frac{2}{3}k, \quad h = k\left(\frac{3 + \sqrt{3}}{3}\right), \quad m = \frac{k}{3},$$

$$b = k\left(\frac{2 + \sqrt{3}}{3}\right), \quad r_2 = \frac{2k(1 + \sqrt{3})}{3}, \quad p = \frac{2 + \sqrt{3}}{5}.$$

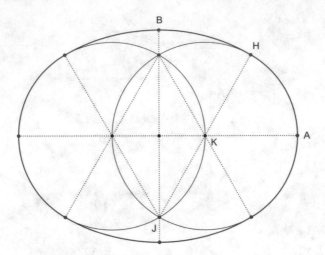

Fig. 6.3 Serlio's fourth construction

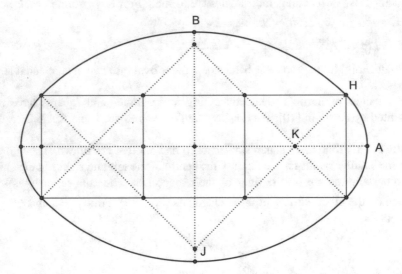

Fig. 6.4 Rosin's three square construction

We come now to a selection from Dotto's work dedicated to Harmonic Ovals [3] and their use in historical buildings. They are ovals where the ratio p is a rational number. In order for this to happen the hypotenuse of the triangle OJK is chosen to be rational. Which is why one starts with a Pythagorean triangle[1] to build a harmonic oval. According to the ratio p these ovals receive a different name, deriving from Pythagorean musical theory. We have chosen two of them, both

---

[1] A right triangle with integer side lengths.

Fig. 6.5 Rosin's four circle
construction

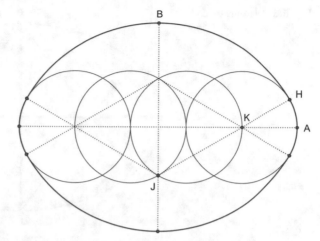

based on the simplest Pythagorean triangle, and decided to keep the Greek names
suggested by the author of the book, which refer to musical intervals.

Many roman amphitheatres were built using a Pythagorean triangle with 3-4-5
proportional side lengths to place the centres of the arcs (see [15]). We will go more
in detail about it in Sect. 8.2, dedicated to the Colosseum.

We start with an oval attributed (see [2, 7]) to Jacopo Barozzi da Vignola.

Diapente (Oval of the Fifth) We start with OKJ as a 3-4-5 triangle, then use a
symmetric copy of it and use the vertex opposite to J as B. We thus have $k, j = b$
$= \frac{4}{3}k$ (Fig. 6.6), which yield

$a = 2k, \; r_1 = k, \; h = \frac{8}{5}k, \; m = \frac{4}{5}k, \; r_2 = \frac{8}{3}k, \; \beta = \arctan\frac{4}{3}, \; p = \frac{2}{3}.$

The next oval enjoys two very special properties.

Diatessaron (Oval of the Fourth) After $j = \frac{4}{3}k$ we take $r_1 = \frac{5}{3}k$; what we get are
the following values for the parameters:

$a = \frac{8}{3}k, \; h = 2k, \; m = \frac{4}{3}k, \; b = 2k, \; r_2 = \frac{10}{3}k, \; \beta = \arctan\frac{4}{3}, \; p = \frac{3}{4}.$

For $k = 3$ (as in Fig. 6.7) all the basic points have integer coordinates, since

$k = 3, \; j = 4, \; a = 8, \; h = 6, \; m = 4, \; b = 6$

and this is also the case with the Diapente oval when $k = 15$, for example, but the
Diatessaron oval delivers the simplest possible combination[2]—it is in some sense a
superharmonic oval. What is even more special, is that in this case we have $j = \hat{j}$, where $\hat{j}$
is given in Chap. 5 by formula (5.3), which means that of all the ovals with axis
proportion $p = \frac{3}{4}$ this is the one with the smallest radius ratio, in some sense the one
closest to a circle.

---

[2]Using a reversed 3-4-5 triangle one can get another integer combination of the same six
parameters two of which are smaller than in the Diatessaron example presented (a=9, b=7,
k=4, j=3, h=8, m=4)

Fig. 6.6 Diapente oval

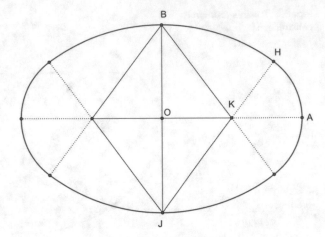

Fig. 6.7 Diatessaron oval

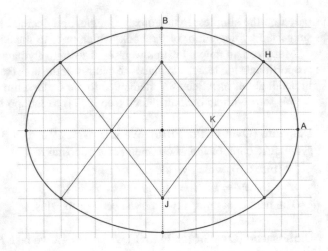

These last two are simple harmonic ovals, but any Pythagorean triangle and an integer (or even rational) value for r_1, will yield a rational value for p, and thus a harmonic oval (see again [3] for comparisons between different harmonic ovals).

The next shape also contains equilateral triangles, and enjoys, as the Diatessaron oval, the minimal ratio property for ovals of the same proportion. It was also quite popular for roman amphitheatres (see [15] where the name was given by Wilson Jones).

Inscribed Equilateral Draw four identical equilateral triangles, two with a common side on the vertical axis, forming a rhombus inscribed in the oval, and two with a common side on the horizontal axis, forming a rhombus with the four vertices as arc centres (see Fig. 6.8). We have then $a = j = b\sqrt{3} = k\sqrt{3}$ yielding

Fig. 6.8 Inscribed
equilateral oval

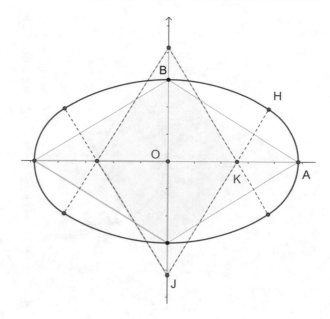

$$h = \frac{1+\sqrt{3}}{2}b, \ m = \frac{3-\sqrt{3}}{2}b, \ r_1 = b(\sqrt{3}-1),$$
$$r_1 = b(\sqrt{3}+1), \ \beta = \frac{\pi}{3}, \ p = \frac{\sqrt{3}}{3}$$

Another famous oval (see again for example [3, 9]) is the one called the Golden
Oval, or Vignola's Golden Oval (see for example [2]) and this is because
$p = \frac{\sqrt{5}-1}{2} = \frac{1}{\Phi}$, where Φ is the famous Golden Ratio.

Vignola's Golden Oval Both A and J are taken at a distance from O equal to twice
the distance $\overline{KO}$ (Fig. 6.9). So $a = j = 2k$ yields
$$r_1 = k, \ h = k\frac{5+\sqrt{5}}{5}, \ m = \frac{2k\sqrt{5}}{5},$$
$$b = k(\sqrt{5}-1), \ \beta = \arctan 2, \ p = \frac{\sqrt{5}-1}{2} = \frac{1}{\Phi}.$$
For the next four ovals we have used the paper [8] by López Mozo as reference.
From the original 1560 table in [11] she deduces the following construction.

Ruiz's Oval As in Serlio's fourth construction there are two equilateral triangles
inside the oval (Fig. 6.10). Draw them with a common side on the horizontal axis
and choose B and J to be the two other opposite vertices. Since $j = b = k\sqrt{3}$ we get
$$a = k(2\sqrt{3}-1), \ h = k\sqrt{3}, \ m = k(3-\sqrt{3}), \ \beta = \frac{\pi}{3}, \ p = \frac{6+\sqrt{3}}{11}.$$
The next construction appeared in Vandelvira's work [14] in 1580 as a drawing
for an arch.

Fig. 6.9 Vignola's golden
oval

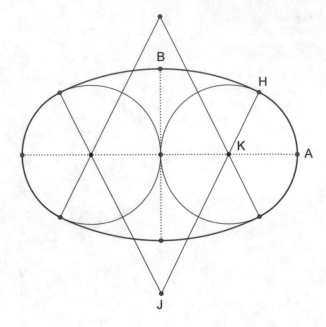

Fig. 6.10 Ruiz's oval

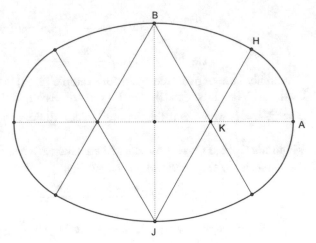

Vandelvira's Oval Two tangent circles are drawn as the ones determining the left
and right arcs, and their centres are chosen as two vertices of two equilateral
triangles (Fig. 6.11). Each third vertex is chosen as the centre of one of the other
arcs. We have $a = 2k$ and $\beta = \frac{\pi}{3}$, yielding

$$r_1 = k, \quad j = k\sqrt{3}, \quad r_2 = 3k, \quad b = k(3 - \sqrt{3}),$$

$$h = \frac{3}{2}k, \quad m = \frac{k\sqrt{3}}{2}, \quad p = \frac{3 - \sqrt{3}}{2}.$$

Fig. 6.11 Vandelvira's oval

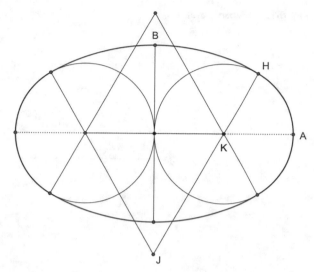

Fig. 6.12 San Nicolás oval

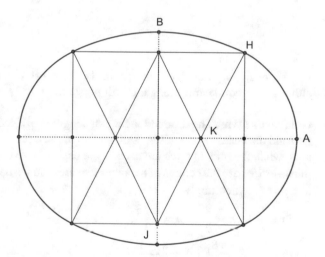

   With the 1639 construction by San Nicolás in [12] we go back to inscribed squares.

San Nicolás Oval A square is divided into two equal rectangles, the centres of which are used as centres for the arcs with a shorter radius. The midpoints of the horizontal sides of the square are used as the other two centres needed (Fig. 6.12). The vertices of the square are chosen as connection points. From $j = 2k$ and $h = 2k$ we get

$$m = 2k, \quad a = k(1 + \sqrt{5}), \quad b = 2k(\sqrt{5} - 1),$$
$$r_1 = k\sqrt{5}, \quad r_2 = 2k\sqrt{5}, \quad \beta = \arctan 2, \quad p = 3 - \sqrt{5}.$$

Fig. 6.13 Gelabert's oval

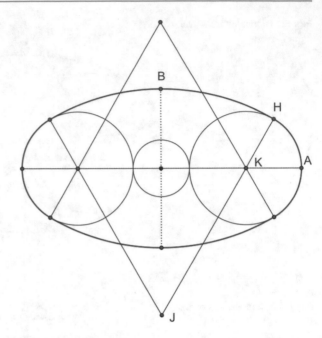

The oval proposed by Gelabert in 1653 in [4] is somewhat more original, although the equilateral triangle is still present.

Gelabert's Oval The desired longer axis length is divided into five parts. The first two and the last two are chosen as diameters of the circles used for the left and right arcs, while the middle three are used as the basis of two equilateral triangles whose third vertices are the centres of the other two arcs. So we have (see Fig. 6.13) $a = \frac{5}{3}k$ and $\beta = \frac{\pi}{3}$, which imply

$$j = k\sqrt{3}, \quad h = \frac{4}{3}k, \quad m = \frac{k\sqrt{3}}{3}, \quad b = k\left(\frac{8 - 3\sqrt{3}}{3}\right),$$

$$r_1 = \frac{2}{3}k, \quad r_2 = \frac{8}{3}k, \quad p = \frac{8 - 3\sqrt{3}}{5}.$$

We end up this list of ovals found in the literature with an oval from an illustration in the Wiener Sammlung, the richest collection of medieval drawings. It can be found in Huerta's paper [6], where the work by Bucher on medieval architectural design methods [1] is mentioned.

Gothic Arch Oval This time the idea is not to draw ovals with forms other than circles inside. Easy-to-use methods were probably more advisable when building an arch, as opposed to forms which had to have some kind of perfection. In any case this oval also features two circles each going through the centre of the other, as in Serlio's fourth oval. The desired longer axis is divided into three parts. The two inner points are chosen as centres for the shorter radius circles (Fig. 6.14); take then

Fig. 6.14 Gothic arch oval

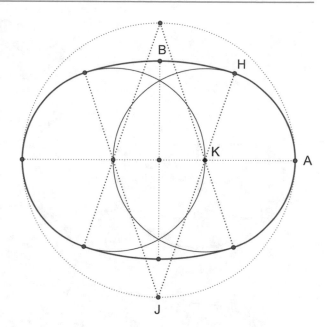

two points at a distance equal to half the longer axis to choose the centres of the remaining arcs. Given a choose $k = \frac{a}{3}$ and $j = a$, we can then calculate

$$h = a\frac{5 + \sqrt{10}}{15}, \quad b = \frac{a(\sqrt{10} - 1)}{3}, \quad m = \frac{a\sqrt{10}}{5},$$

$$r_1 = \frac{2}{3}a, \quad r_2 = a\frac{2 + \sqrt{10}}{3}, \quad \beta = \arctan 3, \quad p = \frac{\sqrt{10} - 1}{3}.$$

Normalized versions (the ovals all have the same value for a) of the preceding ovals are displayed in Fig. 6.15 to enable comparison.

The end of this chapter is dedicated to a construction by the author, which makes use of the properties of ovals with minimum radius ratio, developed at the end of Chap. 5. It is a golden oval in the sense that the Golden Ratio appears in the formulas for the parameters; the longer axis is taken twice the length of the shorter axis.

Having chosen the smaller axis OB one constructs OA with double the size. The procedure now follows the construction at the end of Chap. 5, as in Fig. 5.3, and the result is displayed in Fig. 6.16. Having chosen $a = 2b$ and $\beta = \arctan 2$, the remaining parameters can be calculated via formulas (5.3) and (4.8), the value of the Golden Ratio being $\Phi = \frac{1+\sqrt{5}}{2}$,

$$j = (1 + \Phi)b, \quad k = \frac{(1+\Phi)}{2}b, \quad h = \Phi b, \quad m = (\Phi - 1)b, \quad \text{and} \quad p = \frac{1}{2}.$$

Note that $h = b + m$, that $j = b + h$ and that $k = \frac{b+h}{2}$; this means that another way of drawing this oval is possible, starting with b, then $a = 2b$ and then b 's golden ratio h and m by subtraction of b from h, yielding point H; K is then found considering

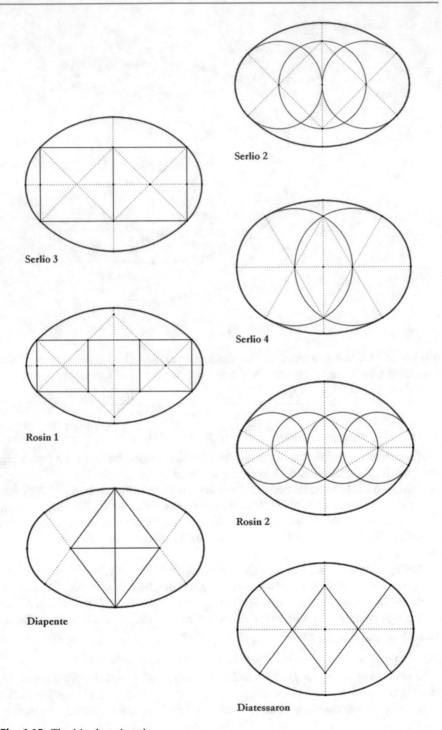

Serlio 2

Serlio 3

Serlio 4

Rosin 1

Rosin 2

Diapente

Diatessaron

Fig. 6.15 The 14 selected ovals

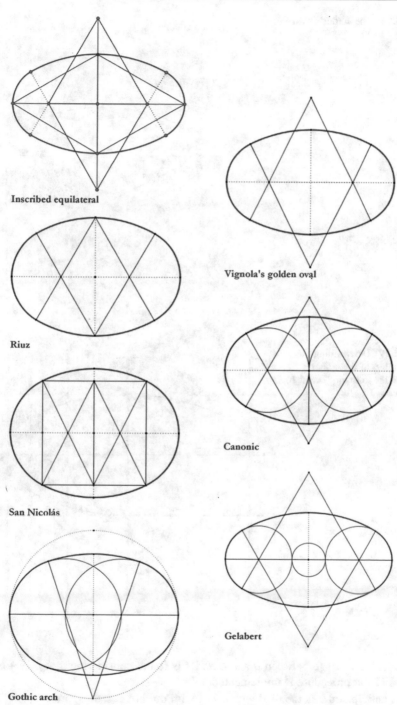

Inscribed equilateral

Vignola's golden oval

Riuz

San Nicolás

Canonic

Gothic arch

Gelabert

Fig. 6.15 (continued)

Fig. 6.16 The author's oval

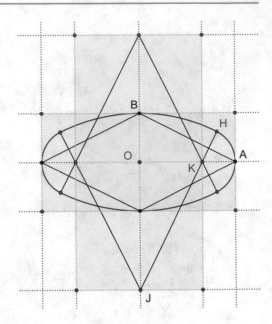

Fig. 6.17 Determining the parameters of the author's oval in the following order: b, a, h, m, k, j

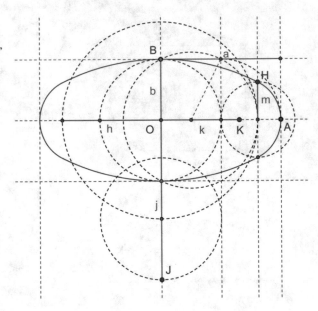

that k is the average between b and h and J is found considering that $j = b + h$. In Fig. 6.17 the procedure is implemented.

As anticipated it is the oval with $p = \frac{1}{2}$ with minimum radius ratio.

References

1. Bucher, F.: Medieval architectural design methods, 800-1560. Gesta. **11**(2), 37–51 (1972)
2. Canciani, M.: Il disegno della cupola di San Carlino alle Quattro Fontane di Borromini: ovale canonico? Disegnarecon. **8**(15), 12.1–12.22 (2015)
3. Dotto, E.: Il Disegno Degli Ovali Armonici. Le Nove Muse, Catania (2002)
4. Gelabert, J.: De l'art de picapedrer. (Facsimile ed. of a 1653 ms in the library of the Consell Insular de Mallorca, Palma de Mallorca). Instituto de Estudios Baleáricos, Palma de Mallorca (1977)
5. Gentil Baldrich, J-M.: La traza oval y la sala capitular de la catedral de Sevilla. Una aproximación geométrica. In: Ruiz de la Rosa, J.A. et al. (eds) Quatro Edificios Sevillanos, pp. 77–147. Colegio Oficial de Arquitectos de Andalucía, Demarc, Occidental, Seville (1996)
6. Huerta, S.: Oval domes, geometry and mechanics. Nexus Netw. J. **9**(2), 211–248 (2007)
7. Kitao, T.K.: Circle and Oval in the Square of Saint Peter's. New York University Press, New York (1974)
8. López Mozo, A.: Oval for any given proportion in architecture: a layout possibly known in the sixteenth century. Nexus Netw. J. **13**(3), 569–597 (2011)
9. Rosin, P.L.: A survey and comparison of traditional piecewise circular approximations to the ellipse. Comput. Aided Geom. Des. **16**(4), 269–286 (1999)
10. Rosin, P.L.: On Serlio's constructions of ovals. Math. Intell. **23**(1), 58–69 (2001)
11. Ruiz, H. el Joven.: Libro de arquitectura. Ms. R.16, Biblioteca de la Escuela Técnica Superior de Arquitectura, Madrid (ca. 1560)
12. San nicolás, Fray L. de.: Arte y uso de architectura. Primera parte (1639). Rpt. Albatros Ediciones, Valencia (1989)
13. Serlio, S.: Tutte l'Opere d'Architettura et Prospettiva di Sebastiano Serlio [...]. G. de' Franceschi, Venezia (1619)
14. Vandelvira, A. de.: Exposición y declaración sobre el tratado de cortes de fábricas que escribió Alonso de Valdeelvira por el excelente e insigne architecto y maestro de architectura don Bartolomé de Sombigo y Salcedo, maestro mayor de la Santa Iglesia de Toledo. Ms. R.10, Biblioteca de la Escuela Técnica Superior de Arquitectura, Madrid (ca. 1580)
15. Wilson Jones, M.: Designing Amphitheatres, vol. 100, pp. 391–442. Mitteilungen des Deutschen Archäologischen Instituts, Römische Abteilung (1993)
16. Zerlenga, O.: La "forma ovata" in architettura. Rappresentazione geometrica. Cuen, Napoli (1997)

Borromini's Ovals in the Dome of San Carlo alle Quattro Fontane in Rome

We have always been surprised, fascinated and intrigued by Borromini's architecture, especially if we consider how carefully his drawings were made, whether they were perspective sketches made with a pen or orthogonal projections made with a pencil, where shapes, dimensions and proportions were displayed. Graphite, "which had been used since the last thirty years of the sixteenth century" (Joseph Connors) is what enabled Borromini to make clear, precise and detailed representations, increasing his control over the project. That is why it makes sense to ask which drawing was the basis for the building of the dome, or at least which elements or data he needed for the purpose. His precision and his obsessive control of the project and of its realisation, as well as his marginal notes on the drawings where he illustrates his lines of reasoning, allow, justify and legitimate this study of ours. Aware that we cannot be totally sure of our results, we can suggest a method which can be the basis for further investigations and new hypotheses.

The complex of San Carlo alle Quattro Fontane, also known as San Carlino, was the first independent work of Francesco Castelli known as Borromini, who had worked until then on the construction sites of Carlo Maderno and Gianlorenzo Bernini in Rome. The architect, who was born in 1599 in Bissone (in the Swiss Canton of Ticino), was given the assignment for the project of the church and for the annexed cloister of the Trinitarians in 1634, and worked on it at various stages until his death in 1667, leaving the façade unfinished; it would be completed in the following ten years by his nephew Bernardo.

Joseph Connors (in [7], contained in [12][1]), one of the most enthusiastic and distinguished researchers of Borromini's work, asserts that in the church of

(Written with *Margherita Caputo*—Freelance architect. Roma, Lecce, Carpignano Salentino (Italy). margherita.caputo@gmail.com)

[1]This is a collection of the researches made in connection of the exhibition "Il giovane Borromini. Dagli esordi a San Carlo alle Quattro Fontane" organised at the Museo Cantonale d'Arte in Lugano from Sept. 5th to Nov. 14th 1999.

© Springer International Publishing AG 2017
A.A. Mazzotti, *All Sides to an Oval*, DOI 10.1007/978-3-319-39375-9_7

San Carlo alle Quattro Fontane we find ourselves "inside an enigma". This enigma has stimulated a number of researchers into investigating various aspects of this work of art, according to their respective interests and competences.[2]

This study aims at giving an answer to the questions which we think have, so far, been left unanswered: how did the project of the dome surmounting the church develop?[3] Which parameters are involved in its shape and in the shape of the coffers at the intrados? Why are there so few drawings of the dome in comparison with those regarding the Trinitarians' complex, and why do none of these answer our questions? Is it enough to assert that the shape of the dome was determined by the models which Borromini used to control the project? How did he work on the building site?

Our starting point is the theory formulated in 1999 by Alessandro Sartor together with Margherita Caputo ([18]) using the data of the 1998 survey; the incomplete results were presented at the 2000 workshop at the Biblioteca Hertziana in Rome and subsequently published in the proceedings of the workshop. The problems that emerged during the research made it necessary to involve a mathematician for a specific contribution on polycentric curves. The theory was then reconsidered together with Angelo A. Mazzotti, after it was realised that a simple and effective procedure[4] for the project and the construction of the dome was not in sight. Even after the main curves for the impost of the dome and the lantern and for the outline of the two vertical sections, had been detected, it wasn't clear how the project and the building had been planned. But Borromini's competence in mastering the oval shape was evident, as suggested by one of the authors in [13].[5] From that time on other researchers have tried to solve the puzzle, either making use of Sartor's data or carrying out surveys with more advanced techniques. But none of them has come, in our opinion, to satisfying conclusions.[6] After 16 years the study which begun in 1999 has been resumed and completed. What follows are the results.

San Carlino's "Fabbrica", like other projects by Borromini, is documented by a significant number of drawings preserved in the Graphische Sammlung Albertina in

---

[2]For references on San Carlo alle Quattro Fontane see [10, 12].

[3]As pointed out by Paola Degni, who has directed the restoration site of the entire complex of the Trinitarians, in different times between 1986 and 2000, it is more precise to call it a vault and not a dome ([9], p. 378); we will carry on calling it a dome as indicated by Borromini himself in his drawings.

[4]It is important to stress that we were not looking for an abstract rule; instead we wanted to simulate the process made by Borromini when trying to control the form by means of the properties of the chosen geometrical shapes.

[5]In the cited paper Mazzotti deduces a value of 1.4461 for the ratio of the impost axes from a preliminary drawing by Caputo. Calculations presented in Subsection 7.2.1 yield the more reliable value of 1.4636.

[6]Notable examples are: McCrossan's dissertation [14], where he examines Borromini's drawings for the church using Sartor's survey, concentrating on the ground plan constructions, and Hill's paper [11], where San Carlino's ground plan is analysed to discover Borromini's ideas for the construction.

Vienna (labelled with the initials AZRom). In many of them[7], mainly concerning the ground plan, the geometrical construction is evident, leading one to think that there are no mysteries to be solved. But these are the drawings, the more clear and readable ones, that Borromini made between 1660 and 1662, many years after the building site had opened, at the time when he was preparing the publication of his "opera omnia"; these drawings are purely theoretical and do not correspond to the present building.[8]

The drawing which, it is agreed, illustrates the genesis of the project for the church is AzRom171 (Fig. 7.1). In this drawing, according to J. Connors (who in [7] dates it back to 1634, with revisions in 1638) at least three different levels can be distinguished, as in an archaeological dig. These correspond to three different possibilities for a ground plan of the church and for the buildings surrounding it. From a central plan church with a hemispherical dome supported by corner pillars, similar to St Peter's in the Vatican, one moves on to a four-leafed clover shape (see again Connors in [7], pp. 478–479), on a plan with four equal arms with deep semicircular apses on two axes and again a hemispherical dome; and finally to the compression of the cross axis to obtain a plan based on an irregular octagon surmounted by a dome having polycentric curves as horizontal sections (that is, made of arcs of a circle sharing a common tangent at the connection points). To sum up we can say that a central-plan church with a hemispherical dome was converted into a rectangular plan church with a polycentric dome (Fig. 7.2). In this respect it is useful here to quote Bellini's interpretation of Borromini's oval, which should not be considered as another regular shape, in the tradition of Peruzzi-Serlio-Vignola, but the result of an empirical distortion of the circle [1]. This supports the idea that Borromini had mastered the use of ovals, allowing him to adapt them to his needs without having to use codified shapes[9] (see the examples on display in Chap. 6).

If we accept this hypothesis and the consequent mixtilinear shape of the ground plan of the church up to the cornice and over the columns which give rhythm to the space, the ceiling over the big arches supporting the polycentric cornice on top of which the dome was to be built remains to be defined (Fig. 7.3). Borromini knew the effect he wanted to create, as can be read in the marginal notes to drawing AZRom218 (which probably refers to San Carlino) where different types of coffered vaults are sketched: "Questo e un modo di/fare alzare et/slontanare più/ le volte con farle/parere più che/non sono alte" ("This is a way to lift and distance vaults, making them look higher than they are" our trans.); a light-weight roofing

---

[7]Drawings AZRom168, 169, 170r, 173, 175, 176, 177. See J. Connors, in [7], p. 478.

[8]Comparison of these drawings with the survey has revealed that a church built according to them would have been larger than the present one.

[9]Bellini's opinion on the subject can be read in [2] p. 28: "Borromini uses geometry as an instrument, his only goal being the visual effect [...] Whenever a conflict arises between architecture and geometry Borromini and his contemporaries always give priority to the former"(our transl).

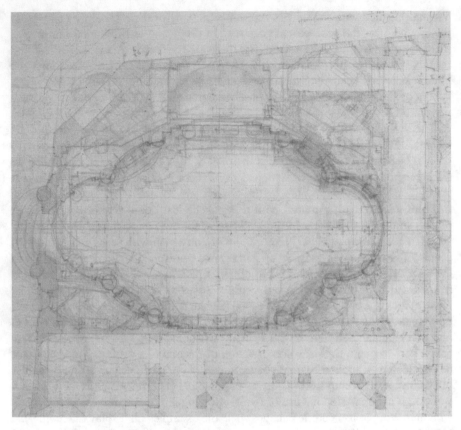

Fig. 7.1 The first drawing for San Carlo alle Quattro Fontane where at least three different solutions can be seen as in a watermark. The compression of the plan along the minor axis is clear. In this image the entrance is to the left and the high altar to the right *(detail of the drawing AZRom171, copyright Graphische Sammlung Albertina, Vienna, with kind permission)*

over the church, making it as luminous and airy as possible (Fig. 7.4). The dome's intrados would not be smooth, or painted, or ribbed in any way. The obvious reference is to the Pantheon in Rome, where, as well as in S. Peter's, "Borromini spent most of his time [. . .] directly contributing to what were considered at the time the two most important architecture textbooks in the world's history" (our transl. from [8], p. 200), but also to domes with a more complex intrados, where the eye, following various pattern lines, is led to believe that what it sees is different than what really is.

7.1   The 1998 Survey

As the 1999 Borromini celebrations approached (400 years since his birth), the Accademia di Architettura di Mendrisio commissioned Alessandro Sartor, professor of Photogrammetry at the Università degli Studi "La Sapienza" of Rome, to

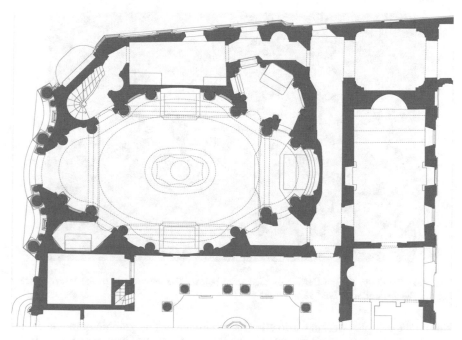

Fig. 7.2 San Carlino's ground plan according to the 1998 survey (drawing by Caputo)

Fig. 7.3 The façade and the inside (at the side of the altar) of the church of San Carlo alle Quattro Fontane, also known as San Carlino *(copyright Sophie Püschmann, with kind permission)*

Fig. 7.4 The dome of San Carlino as seen from below. The entrance is to the left *(copyright Sophie Püschmann, with kind permission)*

survey the complex of San Carlo alle Quattro Fontane. The graphic outcome would then be the base for the 1:1 scale model which the Ticino architect Mario Botta wanted to build on the Lake of Lugano, as a homage to Franceco Borromini, who was born in neighbouring Bissone in 1599, and as a basis for historical studies. They would also be partially used for the restoration by the Soprintendenza ai Beni Architettonici, directed by the architect Paola Degni, which was taking place at the time.

The survey, carried out using stereoscopic photogrammetry—the most advanced technique of the time, as well as the most suitable one for baroque architecture—and integrated by a direct topographic survey, was coordinated by the architects Margherita Caputo and Elena Ippoliti. Other people involved were: architect M. Unali and photographer A. Di Felice for the pictures and the stereo-photogrammetric project; architects L. Bogliolo, P. Ciurluini, V. Cao and junior architect R. D'Eredità for the photogrammetric restitutions, the direct surveys and the 3D elaborations, and surveyor C. Avallone and architect R. Pace for the topographic surveys.

When the graphic restitution process was complete it was essential to compare the survey data with the numerous drawings, many of them by Borromini himself, of the San Carlino building site, in order to understand the genesis of the project and to identify the ones which matched what was actually built.[10]

---

[10]Comparison between the survey and the project drawings has also been carried out by Margherita Caputo for another of Borromini's works: Palazzo Carpegna in Rome, location today of the Accademia Nazionale di San Luca; the corresponding results have been published in [6].

Surprisingly enough, the plan which best matched the survey of the building was the one drafted in AZRom172, dated 1638 and revised in 1641 and in 1660-1662 as specified by Connors (in [7], pp. 479–480) in his fact sheet for the 1999 Mendrisio exhibition, and not those in AZRom168, 169, 170r, 173, 175, 176, 177 where the geometric construction is made evident.

It was also surprising to discover that other drawings could be considered "execution plans" so close were they to the survey, such as AZRom190 (see Fig. 7.7), describing the plan of the lantern[11], AZRom206r, the final drawing of the niche in the transept pillar, AZRom185, illustrating the metal structure supporting the cross at the top of the church[12], and finally AZRom199 showing the façade of the cloister overlooking the garden, corresponding exactly to the present building, with the exception of the middle cornice.

The survey of the dome showed a series of irregularities: for example the intrados appears in its horizontal projection as having an egg-shaped impost rather than an oval one, being somewhat compressed along the longitudinal axis on the side of the altar; moreover its longitudinal axis does not correspond to that of the lantern, which is slightly rotated; finally the four levels of the coffer decoration do not belong to perfectly horizontal planes, probably because of problems that had occurred on such a complicated site and of changes on the way caused by second thoughts or corrections. The study carried out during the restoration pointed out, as revealed by Degni ([9], p. 377), that a "deep and branched" crack crossed the dome along the longitudinal section "extending all the way down to the top [...] of the arcs of the niches" (our translation from the Italian). This was an old crack that had already occurred shortly after the building was completed; moreover Degni reports that during restoration a "*change of mind* of the architect emerged, in the form of a perspective adjustment on top of the vertical arm of the coffer crosses close to the impost, where a bit of mortar was subsequently added, to correct the optical impression of the height, as seen from below" (our translation, [9], p. 380).

In order to understand the three dimensional geometry of the church, software products suitable to represent the complex surfaces were employed, concentrating on the transition between the mixtilinear cornice and the dome, where the arcs around the bowl-shaped coffers support the dome, which rests on the oval cornice.[13] While constructing this virtual model it became clear that a hypothesis on the geometry behind the project of the dome was needed, in order to understand and simulate how it was actually built.

---

[11] Among the drawings of the Albertina collection, AZRom192 is said to correspond to the lantern of San Carlino. This is not the case, in our opinion, since the shape of the oval is somewhat rounder, the displayed spiral staircase could not be built in the actual wall, the double columns between the niches are missing and most of all, the axis of the oval is much longer.

[12] On this sheet one can read: "no(n) si metera vetriata/nel campo della croce/p(er)che la polvere lo im(bratta?)" ("no glass window shall be put inside the cross, because dust would soil it" our translation); this indicates that scenographic details were important to Borromini, but he was ready to give them up if they might cause unwanted effects.

[13] These models have been published in [18] pp. 385–386.

7.2   The Project for the Dome

Before beginning we should illustrate our starting data.

Examination of the vertical sections corresponding to the main axes of the church and of the horizontally projected data points surveyed has identified the sizes of the impost axes and helped to analyse the geometric elements (axes, heights, curves, centres and alignments) necessary to formulate our hypothesis, data partly highlighted by Sartor in 1999 that we here clarify and correct.[14]

We have then studied the horizontal sections of the dome (photographed in Fig. 7.5 together with the arches supporting it) which we believe were used by Borromini to keep control of its shape and to determine the four levels of coffers. We will here describe the oval of the impost, the oval at the base of the lantern, the ovals defining the coffer levels and their position in space. More ovals are visible on the dome, but we consider it plausible that, for example, those in the limited space between the base of the lantern and the end of the decoration, where the dedication to San Carlo Borromeo is written, were outlined on the dome and not drawn and constructed beforehand. Of the chosen seven ovals the largest one was subjected to two unavoidable constraints—the dimensions of the two axes, which derive, as previously illustrated, directly from the measurements of the church—while the others were the architect's choice for the decoration (four levels of coffers, each consisting of eight crosses and eight octagons and a lantern intended to illuminate the dome from the top).

The authors' research of a simple, logical and practical construction corresponding, with a good degree of approximation, to the survey data, has taken a very long time. Many lines of reasoning have followed in the attempt to simulate the artist's procedure, and just as many constructions have been drawn. It is possible that Borromini went along a similar path, in the end choosing—among many simple, logical and practical constructions—the one which best suited his taste, keeping in mind constructive and static requirements.

7.2.1   The Dimensions of the Dome

For a dome with a circular impost it is enough to fix the centre and a single parameter, the radius[15]; for a dome with an oval drum it is necessary to fix two parameters (the lengths of the axes) and a connection point between two consecutive circle arcs. It is then clear that the choice is not only numerical but also

---

[14]A more rigorous examination of the survey revealed that the longitudinal section has a profile made of two circle arcs, forming (part of) an ogival arch, as Bellini (in [1] p. 393) had acutely conjectured.

[15]To keep it simple we imagine domes where the longitudinal section is obtained by rotation of half the drum shape.

Fig. 7.5 A view of the church from below, at the side of the entrance, including part of the dome and the corresponding supporting arcs and pendentives over the mixtilinear cornice *(copyright Sophie Püschmann, with kind permission)*

aesthetic, since different choices of a connection point for the same choice of the oval axes yield different shapes of the oval, making it either more or less flat.

We have seen that it is the ground plan of the church which determines the dimensions of the axes of the dome. Borromini needs to choose an oval inscribed in the rectangle of the impost, with properties helping him to draw the grid of coffers with which he intends to "decorate" the intrados of the dome. He is acquainted with the construction of an oval whose perimeter can be easily divided into equal parts, but he cannot use it because its minor axis exceeds by about 4 Roman palms (3.8285) the minor axis of the needed curve. Nevertheless he decides to use it as reference for the entire project that we are about to describe. It is Serlio's fourth construction (see Fig. 6.3 in Chap. 6). As conjectured for the Colosseum in the next chapter, a special oval shape is chosen with the only purpose of using it as a basis for the development of the actual, more complicated, realization.

From the surveyed data we have derived the axis lengths of the impost oval 11.7871 m and 8.0535 m, corresponding respectively to 52.7623 Roman palms[16] (52 and 3/4) and 36.0497 Roman palms (36 and 1/20), as well as the corresponding height.[17] We draw the corresponding rectangle with those same sides and we draw

---

[16]The value we are using here for the Roman palm is 0.2234 m.

[17]In the drawing AZRom203, ascribed to Borrominni's nephew Bernardo, the indicated measurements in palms of the axes of the "ovato" of the dome are 52 and 1/2 and 35 and 3/4. Both these values differ about 1/4 of a palm (6 cm) from the measured ones. The axes as surveyed by Canciani in [5] are again different. This may have to do with the fact that the height that he uses for the impost (the height of the level containing the centre of the curvature of the vertical section) is 16.38 m from the floor of the church. We maintain that the impost level is about 2 palms under

an oval following Serlio's fourth construction having the major axis $\overline{AC} = 11.7871$. As anticipated, this oval cannot be used by Borromini as oval for the impost, since its minor axis exceeds the value $\overline{BD} = 8.0535$ (see Fig. 7.6).

Nevertheless let us assume that the impost does have exactly the same axes as such an oval and go on considering the eight equal parts into which the perimeter is divided by its intersections with the symmetry axes and by the connection points, as proved in Sect. 4.3. This oval is not, for any given axis measurement, the only one divided into equal parts by its connection points (see again the cited paragraph), but it is the most geometrically elegant one, also a reference to the Trinitarians (the division of the axis into three equal parts, as well as the 120° angles, a third of the perigon angle)[18]. Further divisions of these eight arcs into two equal parts can be obtained using the bisectors of the angles corresponding to such arcs. We now consider the intermediate points, located once more by means of bisectors, of the sixteen obtained so far. These last identified points on the oval determine the vertices of the wedges considered by Borrromini when imagining a coffer decoration for a dome having Serlio's oval as impost. In a different manner to that of a hemispherical dome, the radii do not converge to the centre of the picture (with the exception of those nearer to the minor axis), but rather to points on the major axis having a distance of 1, 2 and 3 Roman palms from it. The complete picture is displayed in Fig. 7.6.

A similar way of constructing radii governing the forms inside an oval can be clearly seen in the drawing where Borromini plans the lantern for the same church, although the ovals here have different shapes than those in Serlio's fourth construction (Fig. 7.7); the axes of the openings and of the wall sections converge to points on the major axis at fixed distances from the minor one.

Still imagining that there is enough space to build a dome using Serlio's fourth oval, we draw the longitudinal section by means of a semi-circle having $\overline{AC}$ as diameter. Our assumption is that this ideal shape of a dome with Serlio's fourth oval as impost was deformed by Borromini to adapt it to the space available and to his ideas for the project. The architect used a deformation module to radially squeeze the base (as required by the project data) and thrust the longitudinal section upwards, as shown in Fig. 7.8. This deformation of the ground plan generates a flatter oval preserving the wedge directions, as we will see later.

The difference between the minor axis in Serlio's oval and the minor axis of the impost oval that needs to be drawn, represented by the segment BM in Fig. 7.9, measuring 0.4287 m—that is 1 Roman palm and 11/12 (precisely 1.9190, nearly 2 palms)—is used as a *deformation module* in the longitudinal section to determine

---

the level where the coffer decoration starts. Canciani chooses some specific sections of the dome and draws the ovals and their most likely constructions, but he does not relate them with one another, not attempting to explain Borromini's choices. Our attempt is to suggest a logical sequence describing the project development. Dimensions of the geometric shapes are thus secondary to the constructions determining them.

[18]We believe that symbolic references, even when obvious, are nonetheless secondary with respect to constructive or formal choices.

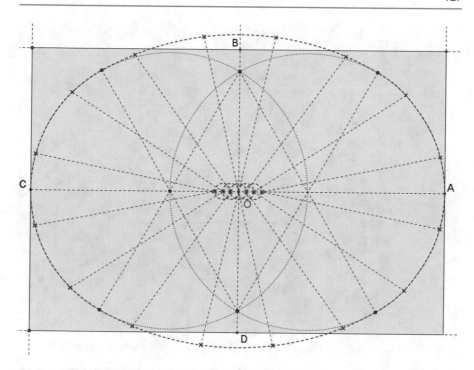

Fig. 7.6 Serlio's fourth construction overlapping the impost rectangle of the dome of San Carlino, and a way of dividing the enclosed area into pseudo-sectors

two new points P and Q having this distance from the centre of the impost. Using them as centres of two circle arcs with radius $\overline{UP}$, each one through the further endpoint of the major axis, one obtains the new profile of the longitudinal section. This *deformation module* calls to mind the idea of *correction* expressed by Simona in [19].

The outline of the resulting longitudinal section is not a single semicircle anymore, it is formed by two symmetric arcs which provide for a leap towards the sky and for an ogival profile diminishing the vertical load exactly on the section where it is greater[19]. This surprising statement is confirmed by the survey which clearly indicates, notwithstanding the asymmetry of the dome previously illustrated, that the project did not include a single semicircle.

7.2.2   The Impost Oval and the Lantern Oval

Regarding the impost, as observed in a paper by one of the authors ([13]), the problem was to draw an oval starting from the dimensions of the two axes, a

---

[19]About the statics of Borromini's domes see [1] (p. 391) or the more detailed [2] (p. 136).

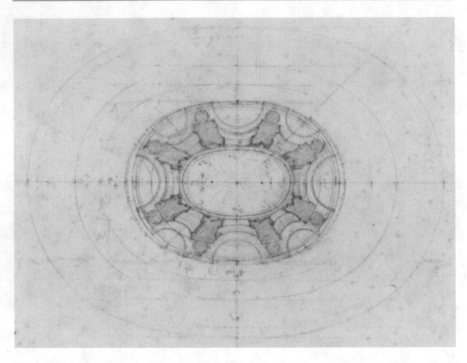

Fig. 7.7   Borromini's drawing for the plan of the lantern. Note the radial directions for the axes of the openings and of the wall sections and where they converge *(detail of AZRom 190, copyright Graphische Sammlung Albertina, Vienna, with kind permission)*

problem which, according to the literature we have access to today, was officially solved some twenty years after work had begun in the church, by the French printmaker Abraham Bosse[20] ([3]). The question is how much was actually known at the beginning of the seventeenth century—although unpublished, or published but not known—and how much Borromini was able to discover by himself. In any case we speculate that he chose as centres on the minor axis line the two points having the distance to the centre equal to half the major axis, proceeding then with one of the two constructions illustrated in Chap. 3 (Constructions 3a and 3b). In this way he was using measurements already present in the drawing and choosing $\overline{OA} = \overline{OJ}$ (i.e. $a = j$ using the notation at the beginning of Chap. 3) as in other oval shapes known at the time, such as Vignola's golden oval, the inscribed equilateral oval and the gothic arch oval, all analysed in Chap. 6.

So having determined J as the intersection between the circle with centre O and radius $\overline{OA}$ and the vertical axis of the impost rectangle, and then having found, for example by means of the Connection Locus[21] method (see Construction 3b in

---

[20]The construction is actually derived from the one appearing in [3], it is Construction 3a in Chap. 3.

[21]First conjectured by Ragazzo in [16], see also Fig. 2.6.

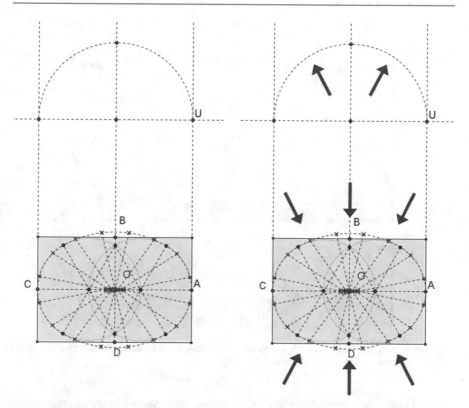

Fig. 7.8 Horizontal and longitudinal sections of an imaginary dome having Serlio's fourth construction as impost oval, and the suggested deformation of it by means of the *deformation module*

Sect. 3.1), a connecting point *H* and a centre *K* on the major axis, the construction of the oval is automatic (Fig. 7.10).

In order to draw the oval for the lantern Borromini could either fix its height, consequently determining the length of its major axis intersecting a horizontal plane with the drawn vertical section profile, or instead choose the major axis making use of elements already in his drawing. The survey shows that the major axis corresponds almost perfectly to the difference of the impost axes values, and this is somewhat confirmed by the notes of Borromini's nephew Bernardo (on AZRom203) where the corresponding sizes verify perfectly this relation: (52 and 1/2) − (35 and 3/4) = (16 and 3/4). Moreover, the axis lengths for the impost in Bernardo's notes differ only by 0.5 and 0.8% from the surveyed values, while the difference between his major axis of the lantern and the one suggested here is minimal (0.835 cm corresponding to 0.04 Roman palms), an "error" of 0.2%. It should also be stressed that in this way Borromini uses a circle which he may have already drawn (if he drew the impost oval by means of the *Connection Locus* method) this time with centre *A* (parts of both circles can be found in

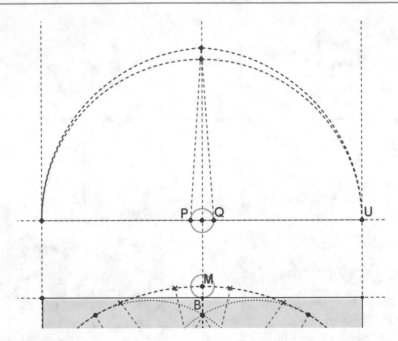

Fig. 7.9 Lifting the contour of the imaginary dome built on the canonic oval by means of the *deformation module*, indicated by the small circle

Fig. 7.11). Supposing this was the actual choice, there were two parameters left to be chosen.

Supported by the surveyed data (we will perform a comparison once the construction is completed), we suggest that two centres may have been chosen on the horizontal axis four Roman palms away from the centre of symmetry—suppose K' is one of them—continuing the sequence used to divide Serlio's oval in pseudo-sectors, also trying to keep the workers' job simple. We also think that the endpoints of the minor axis of the impost were chosen as the centres on the vertical axis. Now to get the connection point H' in the first quadrant one needs to intersect DK' with the circle having centre K' and radius $\overline{A'K'}$ (see again Fig. 7.11). It is interesting to note that the resulting triangle inscribed in the half oval is very close to an equilateral one, since the angle at the vertex A' measures 59.69° (relative error of 0.5%), which suggests that Borromini wanted once more to use the angles contained in the canonic oval.

Having determined the oval for the impost of the lantern, the ground plan now looks like Fig. 7.12. The longitudinal section, after closing it on top with the lantern level, via transportation of the major axis as determined on the ground plan, appears in Fig. 7.13.

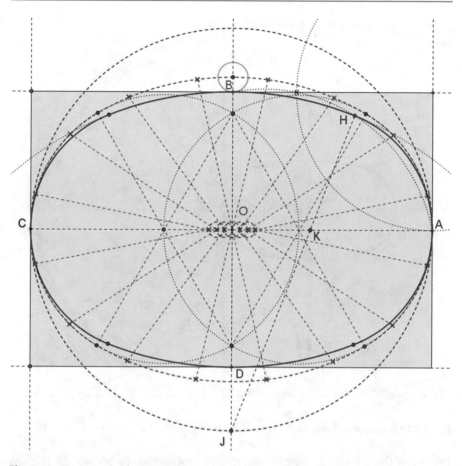

Fig. 7.10 The suggested impost oval

7.2.3   Determining the Height of the Rings of Coffers

Having fixed the two main ovals of the impost and of the lantern, and having drawn
the longitudinal section which is used as a frame for the construction of the dome[22]
Borromini proceeds to determine the levels for the coffering.

The study for the organisation of the coffers is documented in AZRom224
(Fig. 7.14) where the architect draws four rings on five wedges[23]. At the margin
of the drawing he writes: "E si (?) viene meglio in/ quattro (?) / overo in tre": ("And
it is best (done) by four instead of three" our translation). The decision of using four

---

[22]We maintain that the cross section, as well as the other vertical sections through the axes of the
wedges, are a consequence of the construction of the ovals framing the coffer rings.

[23]The same pattern of crosses, octagons and hexagons cam be found as part of the mosaic
decoration of the barrel vault covering the ambulatory of the Santa Costanza mausoleum in Rome.

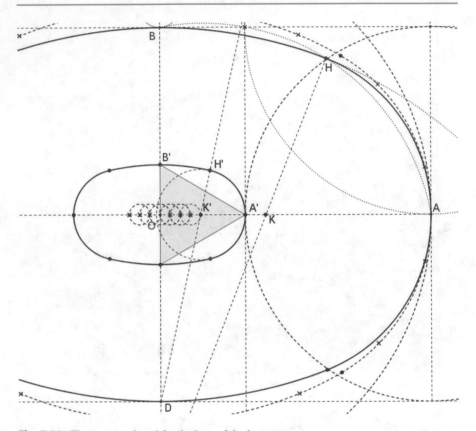

Fig. 7.11 The suggested oval for the base of the lantern.

instead of three levels as initially planned, is probably made because he realizes that, since the first ring is almost vertical, one would anyway appreciate only the top three ones from below, the bottom one being visible only close to the major axis. In this drawing Borromini also indicates the height of the coffers of the bottom ring, 9 palms, equal to the sixteenth part of the length of Serlio's oval.

In order to determine the shape and placement of the four rings of coffers that he had in mind, Borromini uses expedients that allow the coffer decoration to be appreciated from below by modulating their shapes. As recommended in the treatises[24] he leaves some vertical space between the impost and the first level of coffers, choosing the deformation module for this purpose. The height of the first ring, being almost vertical, is then chosen as the width of the central grid square. This corresponds in Fig. 7.15 to choosing the length of the arc WY, which he approximates to the slightly deformed measurement $\overline{RS} = 2.05137$ m,

---

[24]It was common, in hemispherical domes, to start the first coffer ring slightly over the equator level, as recorded nearly two hundred years afterwards by Rondelet in his treaty on the art of building [17].

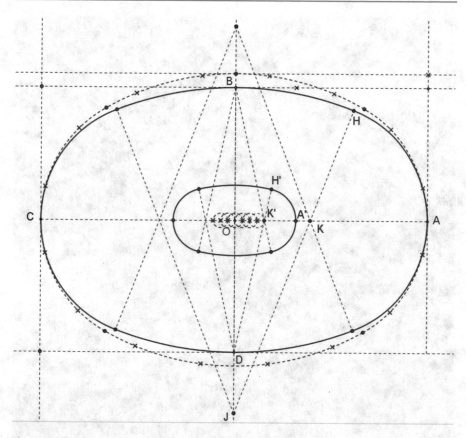

Fig. 7.12 The ground plan with the impost and the lantern ovals.

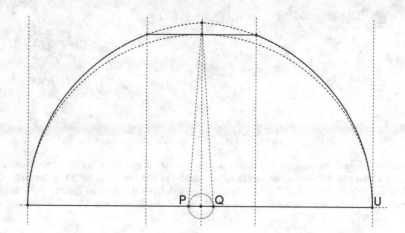

Fig. 7.13 The longitudinal section.

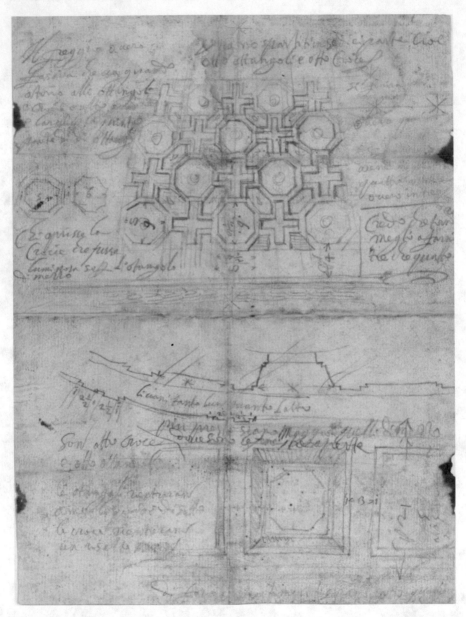

Fig. 7.14 Original drawing by Borromini: at the top the study for the coffer decoration, at the bottom a horizontal section at the height of the middle of the first ring, with the drawing of the window in the axis octagon and the corresponding view from the outside. Notes on the margins illustrate project choices (*AZRom224, copyright Graphische Sammlung Albertina, Vienna, with kind permission*)

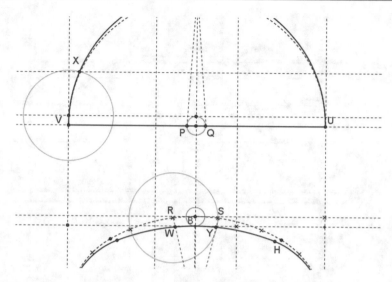

Fig. 7.15 Determining the levels which define the first coffer ring

corresponding to 9 palms and 1/5 (9.1825). This measurement is then modulated by transporting it onto point V, the intersection of the current level with the new profile curve. Point X, obtained by intersecting this circle with the longitudinal section outline, determines the height of the lower band of the coffer decoration, which measures now 1.99697 m, that is nearly 9 Roman palms (8.9390).

The heights of the remaining three rings, which need to leave space for the lantern, are determined by drawing the outline of the central wedge on the longitudinal section, and then proceeding in a similar way, after measuring the width of the central wedge at the current level. As occurs with hemispherical domes, Borromini imagines that the central wedge is seen from the front as an ellipse which is symmetrical with respect to the line of the impost. But in this case, in order to amplify the effect, he chooses as vertex a point which is 0.4287 palms beyond the intersection of the summit of the ogival arch making up the profile, again the deformation module.

To sum up, this ellipse, commonly drawn at the time by means of single points, has $\overline{WY}$ as the minor axis length, a vertex in Z and the base of the longitudinal section as symmetry axis (Fig. 7.16). It is now enough to measure the distance between the two intersections with the height of the last drawn ring, and transfer it onto the profile arc to obtain the next height level—point N—and so on.

It is now possible to compare the above construction with the surveyed data (Fig. 7.17). Note that, because of the asymmetry of the dome as it is today, no construction allows for a perfect match, since we have drawn symmetrical shapes. The variance along the left arch, corresponding to the front of the church (the comparison should me made with the intrados curve), is somewhat compensated by the variance on the opposite side. Note also that the ellipse has been traced with the

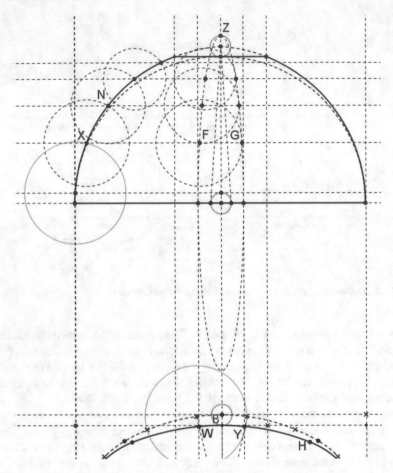

Fig. 7.16 Determining the heights of the remaining levels.

sole purpose of determining the decoration levels; nevertheless its adherence to the middle wedge is quite good.

7.2.4   The Ovals of the Rings of Coffers

Before describing the procedure used to draw the ovals which border the coffer rings, we should remember once again that Borromini was trying to adjust the usual and simpler constructions for domes with a circular drum to his polycentric dome.

In the research that began in 1999 we tried to detect polycentric ovals which could best represent the surveyed data. Because of the asymmetry of the dome, we performed our analysis on an area where the data appeared to be more regular. The result was that the centres did not appear to have a simple arrangement, that the

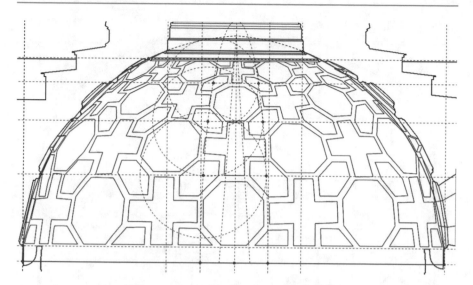

Fig. 7.17 Overlapping of the construction hypothesis and the longitudinal section of the survey

connection points did not lie on the same line, and that the rectangles inscribing the ovals were not similar. We obviously had to look in another direction.

We now project onto the horizontal section the points on the longitudinal section profile corresponding to the different levels, since they determine the length of the major axes of the ovals of the corresponding horizontal sections (Fig. 7.18). Had this been the procedure followed by Borromini so far, he would now be free to choose two more parameters for each oval. Apart from the oval where the coffering starts, basically identical to the impost oval (the bottom part of the dome being, as surveyed, almost vertical), the others were chosen gradually shifting the set of centres on the horizontal axis, using in general a limited number of points on the drawing and keeping a relationship with the original oval at the impost.

We start by using the same centres of the impost oval J, K and their symmetrical points with respect to the axes to draw the oval where the coffering starts, whose major axis is only 6 cm shorter: as can be seen in Fig. 7.19 the two ovals are indistinguishable.

Let's now draw the other ovals bordering the rings of the coffering. Setting I as the centre of the left arc of the impost oval and of the one just drawn (symmetrical to K with respect to the vertical axis), determine four more points on the same axis with distances from it of multiples of one and a half Roman palms towards the centre of the drawing. We now draw the circles having such points as centres through the points projected from the longitudinal section. If we now choose D as the common centre for the circles containing the top-side arcs, it is enough to detect the intersections of the four half lines originating at D through the different centres chosen, each with the corresponding circle, in order to detect the four connection points (see Fig. 7.19). The remaining three connection points for each oval are symmetrical to the one in the top left quadrant with respect to the symmetry axes

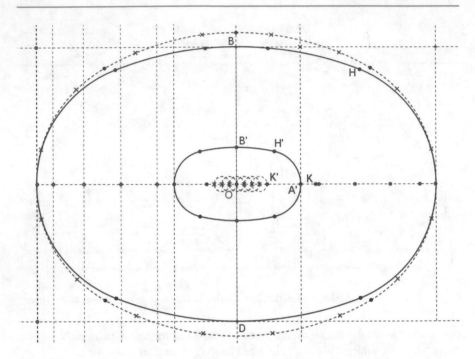

Fig. 7.18 Determining the major axes of the rings of coffering on the ground plan.

and the symmetry centre; through these four points we can now draw the two circle arcs and their symmetrical ones making up each single oval (Fig. 7.20). All of these are constructions of type 6, following the classification in Chap. 3.

Note that the six connection points lying in the same quadrant do not belong to a single line, contrary to the first 1999 hypothesis that they may have been chosen in this way. The two ovals bordering the bottom ring are also very close to each other around the minor axes, causing the corresponding two octagon windows, meant to capture the external light, to be nearly vertical[25]. Finally, note that the most internal oval of the ones just drawn shares only two centres with that of the lantern bottom (points K' and K''), which means that these two ovals are not concentric (see also Sect. 4.4).

We believe that these were the shapes for the frame of the dome, and that the remaining ovals have used them as reference. The dedication to San Carlo Borromeo between the coffer decoration and the beginning of the lantern (in the middle of Fig. 7.4), most likely drawn on the premises, is where Borromini connects the ovals at the end of the decoration and at the start of the lantern in a brilliant way: the margin closer to the lantern is a moulding which follows its oval and is thus concentric to it, while the margin closer to the coffering is concentric to the oval

---

[25]Borromini wanted to use four octagons as windows, but the one on the side of the entrance has always been shut by the façade.

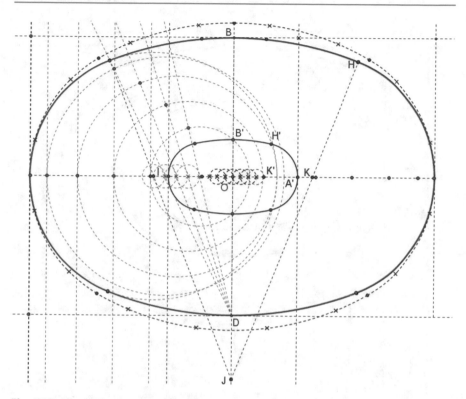

Fig. 7.19 The circles containing the left arcs of the intermediate ovals

ending the coffering. As an example we use points already on the drawing to define the length of the major axes of the two ovals, namely point *I* and point *L* one and a half Roman palms to its right. The result is shown in Fig. 7.21, where the two most internal ovals—sharing common centres K'', *D* and their symmetrical points—and the two next ones—with common centres *T*, *D* and their symmetrical points—are on display. What follows is that the dedication cannot be on a strip of constant width, which is confirmed by the survey.

The ovals drawn to determine the coffering levels, together with the segments subdividing the dome area into wedges in Fig. 7.6 as described at the beginning of Subsection 7.2.1, form a grid defining the space on the dome surface for the main coffers, the ones including either a cross or an octagon. In Fig. 7.22 these lines overlap the ground plan of the 1998 survey. In this picture, as in Fig. 7.17, one can detect the asymmetry of the survey ground plan, which doesn't allow for a perfect match with any project hypothesis. The area where the least adherence is evident is the one around the dedication strip, which, as previously stressed, is the area of greatest distortion in the dome.

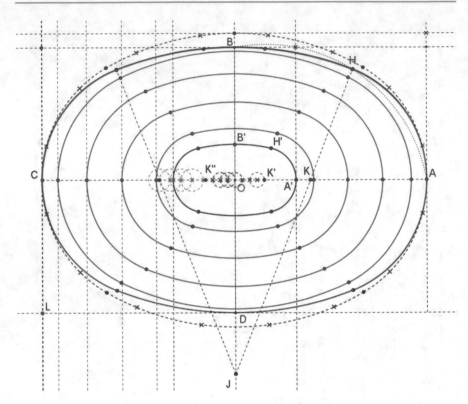

Fig. 7.20 The ground plan of the dome, including the impost oval, the lantern oval and the five ovals defining the coffering

7.3   Conclusions

This study would not be valid, had it not been founded on the carefully analysed data of a scientific survey. Validation has come from Borromini's drawings, where constructions and texts have been detected and interpreted, directing our way. Constantly finding clues for the interpretation of the drawings, by analysing the survey, we have finally identified the geometric cornerstones responsible for the transformation of an idea into a building, by an architect who was a modern artist working on a scientific basis ([15] p. 443).

We maintain that the theory presented here is a convincing answer to the questions posed at the beginning, the sequence of logical steps being compatible with Borromini's way of proceeding. These can be summed up as follows:

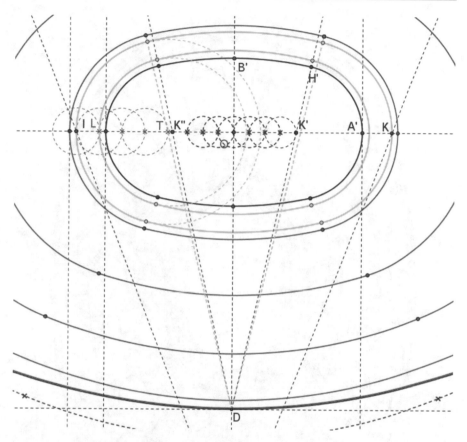

Fig. 7.21 A suggestion for the ovals enclosing the dedication

1. determination of the project data on the horizontal section at the impost height of the dome and on the longitudinal section;
2. drawing of the canonic oval, whose points and properties are going to be useful throughout the procedure;
3. drawing of the impost oval using Bosse's method (or another one) having shortened the minor axis by roughly two palms and having moved the centres lying on it outside of it, thus *deforming* the canonic oval;
4. drawing of the longitudinal section, also deformed in order to obtain an ogival outline. This correction also corresponds roughly to two palms;
5. determination of the lantern oval, probably choosing for the major axis the difference of the impost axes; the centres on the major axis are then chosen at a distance of four palms from the symmetry centre, while for the other two the endpoints of the minor axis of the impost, already on the drawing, are used. Such an oval inscribes a pair of nearly equilateral triangles;
6. definition of the levels of the coffering rings using the width of the central wedge, thought of as an ellipse in its projection onto the longitudinal section.

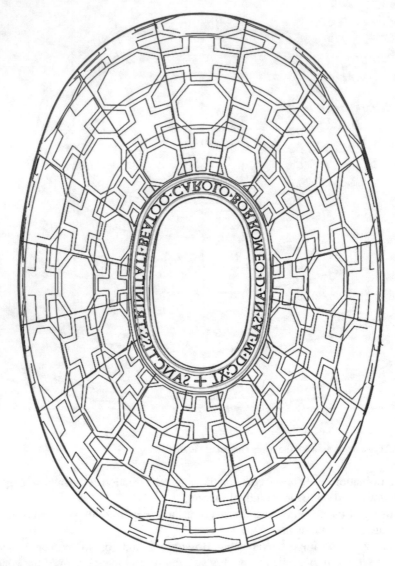

Fig. 7.22 Overlap of the authors' project hypothesis and the 1998 survey. The top part is the side of the entrance

The corresponding heights determine the lengths of the major axes of the ovals outlining the border between the coffer rings;

7. choice of a criterion to draw the ovals of the coffer rings using the major axes and other known points or fixed distances from them, without imposing the measurements of the minor axes. We observe that, differently from what has been suggested in [4], the dome surface cannot be represented by a mathematical function unless one is ready for either very complicated forms or excessive approximations, and we are convinced that art is never the consequence of a straightforward algorithm ([2] p. 100).

We can thus imagine that Borromini prepared on his building site:

1. the drawing of the plan at the impost level including the ovals and the wedges, as in Fig. 7.22;
2. timbers for the centring corresponding to the major axis according to the profile in Fig. 7.13;
3. seven wooden oval shapes as in Fig. 7.22—corresponding to the intrados of the coffers—at levels determined in Fig. 7.17;
4. possibly the timbers for the centring determining the wedges of the coffering, in order to create the grid where the octagons and the crosses were meant to be. The ability of the workforce and the organisation of the site might however have permitted the drawing of the coffers avoiding these;

It is likely that to make things cheaper and easier because of the limited space inside the dome, Borromini organised the timber of points 3) and 4) for only half or a quarter of the dome. In this case we would have in a partial collapse of such a structure a possible reason for the irregularities on the dome surface. This hypothesis does not exclude that Borromini may have produced a model for the three-dimensional check, but we think that it is likely that he made a drawing containing the data of Fig. 7.22, similar to the one for the lantern in Fig. 7.7.

Our theory agrees with some known aspects of Borromini's architectural culture: the union between theory and practice, the knowledge of the rules and practice on building sites and the use of geometry to turn an idea into a project and a form. In this respect we find what Virgilio Spada[26] writes about Borromini, in his letter to cardinal Cesare Rasponi in 1656 very much appropriate:

"Nella scrittura del Cav.re Borromino [. . .]
la melodia delle voci è appoggiata a' numeri;
come in misura la bellezza delle fabbriche [. . .] pretende,
nascer parimenti da numeri,
e che tutte le parti habbino una tale proportione,
che un[a sola] apertura di compasso—senza mai muoverlo—le misuri tutte".
("In Cavalier Borromini's writting [. . .]
the melody of the voices leans on numbers;
as beauty in buildings [. . .] is expected,
born at the same way to numbers,
and all parts should have such proportions,
that a single compass opening
-without moving it- could measure them all." our transl.)

---

[26]Document cited by Tabarrini in [20] (p. 115). Virgilio Spada (Brisighella 1596—Roma 1662), non-professional architect, from 1622 in the Oratorio romano of S. Filippo Neri, played a decisive role in its construction, along with P. Marucelli and, after 1637, with F. Borromini.

References

1. Bellini, F.: La statica delle cupole Borrominiane. Suggestioni dall'antichità e tecniche moderne. In: Frommel, C.L., Sladek, E. (eds.) Francesco Borromini. Atti del convegno internazionale. Roma 13–15 gennaio, pp. 390–405. Electa, Milano (2000)
2. Bellini, F.: Le cupole di Borromini. La "scienza" costruttiva in età barocca. Electa, Milano (2004)
3. Bosse, A.: Traité des Geometrales et Perspectives Enseignées dans l'Academie Royale de la Peinture et Sculpture. L'Auteur, Paris (1655)
4. Canciani, M., Falcolini, C., Spadafora, G.: From complexity to geometrical rule. The case study of the dome of San Carlino alle Quattro Fontane in Rome. In: Gambardella, C. (ed.) X Forum internazionale di studi Le vie dei Mercanti, Napoli. La Scuola di Pitagora, Napoli (2012)
5. Canciani, M.: Il disegno della cupola di San Carlino alle Quattro Fontane di Borromini: ovale canonico? Disegnarecon. **8**(15), 12.1–12.22 (2015)
6. Caputo, M.: Il rilievo di Palazzo Carpegna. In: Atti 2011–2012, pp. 181–207. Accademia di San Luca, Roma (2013)
7. Connors, J.: Un teorema sacro: San Carlo alle Quattro Fontane. In: Kahn-Rossi, M., Franciolli, M. (eds.) Il giovane Borromini dagli esordi a San Carlo alle Quattro Fontane, pp. 459–474. Skira, Milano (1999)
8. Curcio, G.: "Veramente si possono gloriare d'havere sì valenthuomini". I maestri dei laghi e Francesco Borromini tra Corporazioni e Accademia in Roma all'inizio del Seicento. In: Kahn-Rossi, M., Franciolli, M. (eds.) Il giovane Borromini dagli esordi a San Carlo alle Quattro Fontane, pp. 187–208. Skira, Milano (1999)
9. Degni, P.: San Carlino alle Quattro Fontane. Annotazioni sui restauri eseguiti e in corso. In: Frommel, C.L., Sladek, E. (eds.) Francesco Borromini. Atti del convegno internazionale. Roma 13–15 gennaio, pp. 372–380. Electa, Milano (2000)
10. Frommel, C.L., Sladek, E.: Francesco Borromini. Atti del convegno internazionale. Roma 13–15 gennaio. Electa, Milano (2000)
11. Hill, M.: Practical and symbolic geometry in Borromini's San Carlo alle Quattro Fontane. J. Soc. Archit. Hist. **72**(4), 555–583 (2013)
12. Kahn-Rossi, M., Franciolli, M. (eds.): Il giovane Borromini dagli esordi a San Carlo alle Quattro Fontane. Skira, Milano (1999)
13. Mazzotti, A.A.: What Borromini might have known about ovals. Ruler and compass constructions. Nexus Netw. J. **16**(2), 389–415 (2014)
14. McCrossan, J.: Complexity and Simplicity in the plan of San Carlo alle Quattro Fontane (B.A. dissertation, University of Dublin, Trinity College, Department of the History of Art and Architecture) (2008)
15. Oechslin, W.: "Doctrina & Veritas" e prassi: esperienze milanesi di Borromini. In: Kahn-Rossi, M., Franciolli, M. (eds.) Il giovane Borromini dagli esordi a San Carlo alle Quattro Fontane, pp. 437–451. Skira, Milano (1999)
16. Ragazzo, F.: Geometria delle figure ovoidali. Disegnare idee immagini. **VI**(11), 17–24 (1995)
17. Rondelet, J.-B.: Trattato teorico e pratico dell'arte di edificare—vol. 2. (Italian translation of the original French edition and notes by Basilio Soresina). Fratelli Negretti, Mantova (1832)
18. Sartor, A.: Il rilievo della fabbrica di San Carlo alle Quattro Fontane. Un contributo alla conoscenza delle idee progettuali dello spazio interno. In: Frommel, C.L., Sladek, E. (eds.) Francesco Borromini. Atti del convegno internazionale. Roma 13–15 gennaio, pp. 381–389. Electa, Milano (2000)
19. Simona, M.: Ovals in Borromini's geometry. In: Emmer, M. (ed.) Mathematics and Culture II, pp. 45–52. Springer, Berlin-Heidelberg (2005)
20. Tabarrini, M.: Le scale coclidi di Borromini. In: Borromini e gli Spada. Un palazzo e la committenza di una grande famiglia nella Roma barocca, pp. 79–121. Gangemi Editore, Roma (2008)

Ovals with 4n Centres and the Ground Plan of the Colosseum

8

Ovals must have a number of centres which are a multiple of 4. Ovals with more than four centres have been used to align important points in a building and/or to make the oval look more like an ellipse. In this chapter we will extend a property presented in Chap. 2 which allows solving some construction problems for ovals with more than four centres, and illustrate a procedure to draw eight-centre ovals once two of the three radii are known. We will then illustrate the procedure that, according to Trevisan, may have been adopted for the ground plan of the Colosseum, where one four-centre oval and more eight-centre ovals were used.

8.1 The Construction of Ovals with 4n Centres

When the axis lines are given the number of parameters needed to draw an oval with $4n$ centres is $1 + 2n$. For an eight-centre oval (see Fig. 8.1) this may mean closing a quarter oval for which an arc has already been drawn, that is by adding two more arcs to reach the vertical axis. In this case four parameters are given: $\overline{OA}, \overline{OB}, \overline{OC_1}$ and one of the distances of H from one of the axes—the second one being automatically deduced by the fact that H belongs to the circle with centre C_1 and radius $\overline{C_1A}$. In [9]—following an original idea by Ragazzo in [12]—the *Connection Locus* (CL) has proved to work for egg-shapes (closed convex polycentric curves with one symmetry axis) and for generic polycentric curves, when it comes to joining two points with two arcs of a circle imposing the tangents at these two points. What one draws is called a *system of tangents*, which is what one needs to draw the CL where the meeting point of the two missing arcs must be chosen. The infinite choices account for the fifth degree of freedom of eight-centre ovals.

Following [9] we draw the tangent and the normal to the existing arc at point H_1 and call W and X their intersections respectively with the parallel and the perpendicular to AC_1 through B. Find point Y on segment BW such that $\overline{WY} = \overline{WH_1}$. The arc YH_1 of the circle through H_1, Y and B is the CL for the two missing arcs. That is,

© Springer International Publishing AG 2017

A.A. Mazzotti, *All Sides to an Oval*, DOI 10.1007/978-3-319-39375-9_8

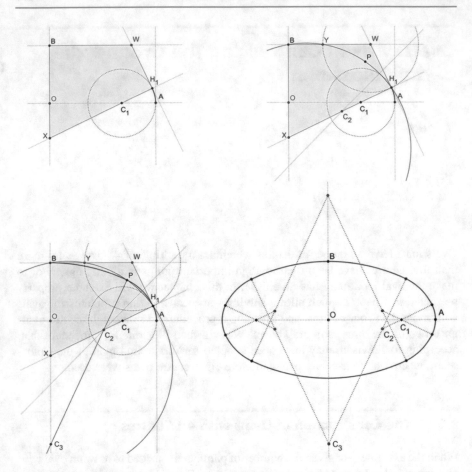

Fig. 8.1 Closing up an eight-centre oval using the CL

for any chosen point on it there exist a couple of arcs which can be used to create a quarter oval. After choosing such a point P we find the second centre C_2 by intersecting the axes of segment PH_1 with the line through H_1 and C_1. Then we draw the line PC_2 to find the next centre C_3 at the intersection with the line of the shorter axis. Arc PB with centre C_3 completes the quarter oval. The last section of Fig. 8.1 shows the complete eight-centre oval obtained, and the eight centres.

This technique can be used whenever two consecutive arcs of any kind of oval are missing.

Suppose instead that one has the two axis measures and the radius measures r_1 and r_3 of the first and third arc, and one needs to know if and where a third arc can be inserted and smoothly connected to the two. We are in a situation described on the left hand sife of Fig. 8.2: looking for a point F such that there must be H and I, respectively on FC_1 and on FC_3, for which $\overline{FI} = \overline{FH}$. As both Rosin in [13] and Trevisan in [15] remarked, such a point lies on the ellipse having foci in C_1 and C_2,

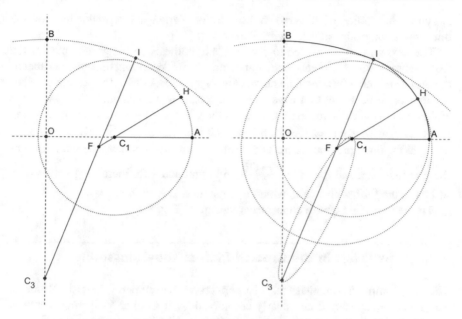

Fig. 8.2 Conditions for point F and the quarter oval deriving from a choice of F on the ellipse

as can easily be shown by observing that since $\overline{FH} = \overline{FC_1} + r_1$ and $\overline{FI} = r_3 - \overline{FC_3}$, this yields

$$\overline{FC_1} + \overline{FC_3} = r_3 - r_1,$$

which is the expected equation. Having chosen point F on this ellipse the rest is easy (see the right hand side of Fig. 8.2).

Not all points on this ellipse are feasible centres for the intermediate arc. They also have to be inside the OC_1C_3 triangle.

This second kind of problem is also a very special case of those investigated by Ragazzo in [12], where a complete set of constructions of general polycentric curves can be found.

Ovals with more than four centres have been widely used: they can be so close to ellipses that many researchers argue whether famous buildings were in fact designed (and/or built) either as eight (or more) -centre ovals or as ellipses: within the common work on the Colosseum [1], Trevisan (see [15]) compares the ground plan of various Roman amphitheatres and discusses the four- and eight-centre options, while Benedetti (see [2]) suggests an eight centre oval construction for Antonio da Sangallo's Vatican dome, as also reported by Migliari in [10], where in general the oval vs ellipse dispute is discussed. Due probably to technical advantages, arches and bridges have often been built using half ovals with 5, 7, 9 or more centres (the formula for a half oval is $2n+1$ centres): see for example

Breymann's studies of lowered arches [3] or Zerlenga's reproductions of the building of the bridge at Neuilly by Perronet in [17].

The next section is devoted to Trevisan's hypothesis on the ground plan of the Colosseum. A series of eight-centre concentric ovals describing the outer perimeter, the arena and other intermediate curves are suggested. As with four-centre concentric ovals (see Sect. 4.4) to draw an oval concentric to a given one—that is with the eight arcs having the same eight centres as the given oval—it is enough to take any number r (positive or negative) such that $r > k - a$, where $a = \overline{OA}$ and $k = \overline{OC_1}$, and add it to the three previous radii. The resulting oval will not have the same form as the previous one, and the ratio $p = \dfrac{\overline{OB}}{\overline{OA}}$ will approach 1 for increasing values of r, just as in the four-centre case, which is illustrated in Sect. 4.4.

To see four eight-centre concentric ovals go to Fig. 8.12.

8.2   The Ovals in the Ground Plan of the Colosseum

The huge amphitheatre that was inaugurated by the emperor Titus in Rome in 80 AD had been started and nearly completed by his father Vespasian, between 69 and 79 AD. Titus' younger brother Domitian, who became emperor in 81 AD, completed the building. The commitment of this family of emperors, the Flavian dynasty, to the construction of this home of entertainment for the people of ancient Rome, gave it the name of Flavian Amphitheatre (Latin: Amphitheatrum Flavium). The name by which it is known today—the Colosseum (or Coliseum)—may have originated from a nearby colossal statue of the emperor Nero, or may simply refer to the capacity and dimensions: it is nearly 50 m high, 160 m wide and 190 m long, and could host tens of thousands of spectators.

After nearly 2000 years most of the Colosseum is still standing (see Fig. 8.3), having survived earthquakes and neglect (see [5] for a study of the effects of earthquakes on the Colosseum), and it was possible, by means of a survey made by the *Department of Representation and Survey* of the university "La Sapienza" of Rome and described in [11], to derive precise information on the points composing the basic shapes of the ground plan of the amphitheatre.

8.2.1   References and Data Used: The Two-Step Method

The data collected in the above cited survey have been used to formulate hypotheses on the design of the Colosseum, many of them contained in the collective work [1] published in 1999. Of the papers included in it we will refer to [15, 14], and [4] respectively by Trevisan, Sciacchitano and Casale, who have used the same data although with different mathematical and computational approaches. In particular we will reproduce the construction described by Trevisan in [15]—who cites, among others, the works by Wilson Jones [16] and Golvin [8]— where he considers the possibility that the architects used, wherever possible,

Fig. 8.3   The Colosseum in 2016 *(copyright Sophie Püschmann, with kind permission)*

integer values of the Roman foot (Latin: *Pes*), one of the units of measure used at
the time.

The perimeter of the Colosseum consists of three stories surmounted by an attic,
forming a closed ring. Each of the three storeys is made up of 80 arcades, the ones
on the ground floor used as entrances for the spectators. The seating inside was
organised on different levels, the lowest rows being reserved for the highest ranks in
society. The ambulatories were curved passages circling the arena at different
distances and levels. Figure 8.4 offers a view of the inside of the building as it is
now.

Following [15], where [16] is cited as inspiration, we consider a two-step
procedure supposedly carried out by the architects of the Colosseum when they
decided what the ground plan of the building should look like. The first step was to
decide which four-centre oval should be considered in order to plan the division of
the perimeter into 80 equal arcades (we have seen in Chap. 7 that a similar idea was
probably used by Borromini—sixteen centuries later—for his plan of the church of
San Carlo alle Quattro Fontane, also in Rome). The second step was to find an
eight-centre oval as close as possible to the first one (or an ellipse, as some
researchers speculate) for the actual perimeter of the amphitheatre and then draw
smaller ones to define the ambulatories and the arena.

8.2.2   The Four-Centre Oval Guideline

As was often the case (see for example [16]) a right-angled triangle with
proportions 3, 4 and 5 was chosen as the base for the construction, as in the case

Fig. 8.4 A view of the inside of the Colosseum *(copyright Gabriel Püschmann, with kind permission)*

of the simplest harmonic ovals described in [7]—two of which have been presented in Chap. 6. The one for the Colosseum is assumed to have sides measuring 81, 108 and 135 Roman feet (in [15] a Roman foot/centimetre conversion coefficient of 29.673 is used). Having chosen the centre of the building and the two symmetry axes, four such triangles are drawn (see Fig. 8.5). Prolongation of the hypotenuses yields the four sectors XCY', XDX', $X'C'Y$ and $YD'Y'$. We then draw the oval with centres C, D', C' and D having a full major axis of 636 Roman feet, which we imagine as a guideline for this stage of the plan (Construction 6 in Chap. 3 can be used). The parameters of this oval, using the same notation as in Chap. 4, are the following:

$$a = \overline{AO} = 318, \quad j = \overline{OD} = 108 \quad \text{and} \quad k = \overline{OC} = 81$$

which yield for the remaining ones (by means of the formulas from Case 6 in Chap. 4)

$$b = 264, \quad h = 223.2 \quad \text{and} \quad m = 189.6,$$

and, by means of formula (4.1),

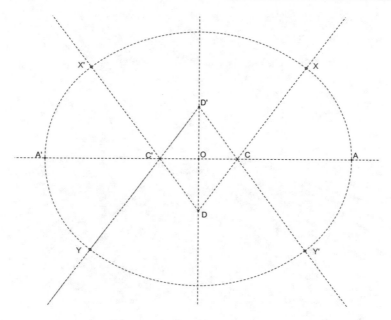

Fig. 8.5 The imaginary four-centre oval used as a guideline for the radial components of the Colosseum and for the actual perimeter form

$$r_1 = 237, \quad r_2 = 372, \quad \beta = arctg\frac{4}{3}, \quad p = \frac{44}{53}.$$

According to Trevisan, this choice of shape had a certain number of advantages, one of which was the possibility of calculating the perimeter with a good deal of precision.

The formula proposed in Sect. 4.3 could be used, since for π the value of $\frac{22}{7}$ suggested by Wilson Jones in [16] is a very good approximation, and since the angle β and its complementary are also well approximated by fractions of the right angle, namely $\frac{13}{22}$ and $\frac{9}{22}$.

At this point of the process, the arcade directions were obtained by the division of angles $X\widehat{D}X'$ and $Y\widehat{D}'Y'$ into 21 equal parts and of angles $X\widehat{C}Y'$ and $Y\widehat{C}'X'$ into 19 equal parts, making a total of 80 rays which are supposed to have been the axes of the pillars of the arcades (with the exception of the four main entrances, which were intended to be wider than the others). See Fig. 4 in [14] for a geometrical reconstruction based on the actual data. Casale in [4] suggests that Ippia's trisector may have been used for the purpose. The result is shown in Fig. 8.6.

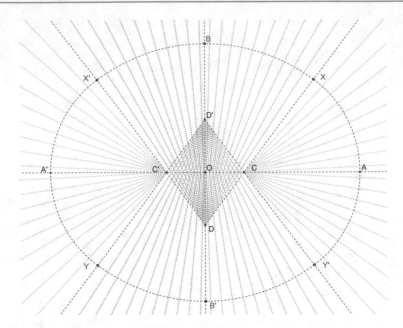

Fig. 8.6 The 80 radial divisions used to define the arcades

8.2.3 From a Four-Centre to an Eight-Centre Oval

The problem with the four-centre oval used to define the directions of the arcades is that, as researchers today agree, its form does not fit the data of the survey. The relatively abrupt way the curve changes its curvature at point X is not what the interpolated curves based on the recorded points show. Other curves close to the four-centre oval—therefore with a value very close to the computed measure of the perimeter—are either an oval with more centres (8, 12 etc. ...) or *the* ellipse with the same axis measures. Following the guidelines in [15] the eight-centre oval hypothesis is described, although there are some researchers who consider the ellipse hypothesis plausible. Migliari's paper [10] on the ellipse-oval dilemma is worth reading.

So, still making use of the points found in the four-centre oval construction, we keep D as a centre for an arc and we discard C using instead a point E chosen 9 Roman feet to the right of C. For two of the three radii we have (see Fig. 8.7)

$$r_1 = \overline{AE} = \overline{AO} - (\overline{OC} + \overline{OE}) = 228 \text{ and } r_2 = \overline{BD} = 372; \qquad (8.1)$$

as observed in Sect. 8.1 (see Fig. 8.2), the third centre has to lie on the ellipse which is the locus of the points having the sum of their distances to D and E equal to $r_2 - r_1 = 144$. Such an ellipse runs through C, since $\overline{CD} + \overline{CE} = 144$. Of the different choices for the third centre G, at the end of his paper [15] Trevisan suggests a feasible point yielding another 3-4-5 triangle, similar to the one used

Fig. 8.7 Choosing two of the
three centres of the eight-
centre quarter-oval

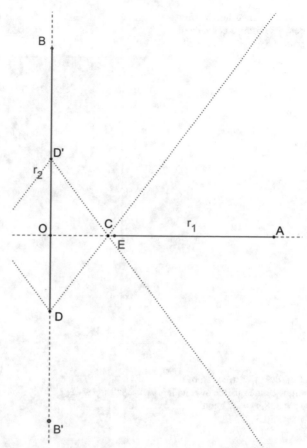

for step 1 of the plan, namely point L inside OD such that $\overline{OL} = 67.5$, forming the
3-4-5 triangle OLE with sides 67.5, 90 and 112.5. Point G is now the intersection of
LE with the above described ellipse (see Fig. 8.8). This choice leads to what
Trevisan calls the "ideal configuration" (our translation) of the whole setup of
eight-centre ovals describing the perimeters around and inside the Colosseum.
The drawing of the corresponding quarter oval is now straightforward. An arc
with centre E and radius $r_1 = \overline{EA}$ is drawn up to the intersection I with the line
GE, yielding for the intermediate radius the value $r_3 = 350.5$. A second arc with
centre G and radius $r_3 = \overline{GI}$ is drawn up to the intersection with DG, yielding the
point J. The third arc with centre D and radius r_2 is part of the original four-centre
oval (see Fig. 8.9).

It is suggested that the choice of this oval (or of an ellipse as some believe)
instead of the 4-centre oval originated in the desire to have a rounder form halfway
around each quadrant and at the same time a very close value to the already
calculated one for the outer perimeter of the building. The resulting shape is
represented in Fig. 8.10, together with the previously drawn four-centre oval.

Fig. 8.8 Choosing the third
centre G using a convenient
point L

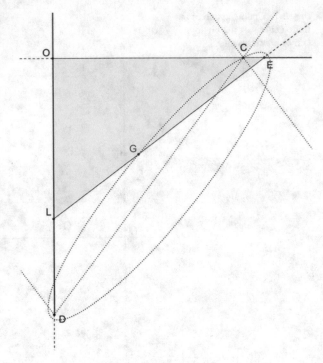

Fig. 8.9 Construction of the
eight-centre quarter oval with
$r_1 = 228$, $r_2 = 372$ and
$r_3 = 350.5$

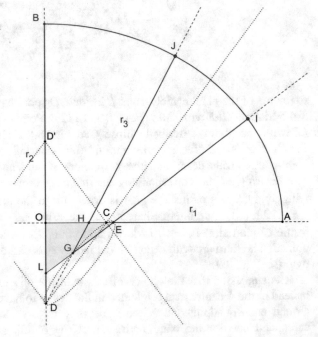

Fig. 8.10 The resulting
eight-centre oval (in *blue*) and
the previously constructed
four-centre oval (in *red*)

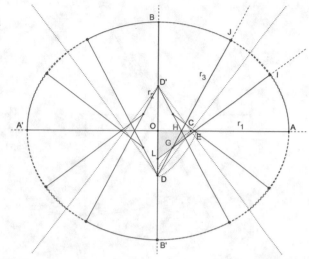

It is then assumed that all the other closed curves belonging to the horizontal
sections (for example those defining the ambulatories) are eight-centre ovals
sharing the same centres as the oval describing the outer perimeter. As explained
in Sect. 8.1, where Sect. 4.4 is cited as reference, it is enough to choose a common
value to add to or subtract from the three radii to define another oval, which as
explained will not share the same shape with the others. Trevisan suggests that
multiples of the number 27 were used to decide where these ovals should lie,
observing that this rule could be used on both axes still using the same centres.

Choosing for example 135 Roman feet as half axis $\overline{ON}$ for the inner oval (see
Fig. 8.11), the one defining the arena (for which unfortunately no data could be
collected), the radius $\overline{EN}$ is 45, that is 183 shorter than the corresponding radius of
the outer perimeter oval. Subtraction of the same number from the other two radii of
the outer oval (see (8.1)) one gets the values 167.5 and 189, the latter value
indicating that $\overline{OR}$ has to lie on a circle with centre O and radius $27 \times 3 = 81$, since

$$\overline{OR} = \overline{RD} - \overline{OD} = 189 - 108 = 81.$$

The resulting oval is represented in Fig. 8.11, where concentric circles of radii
27, 54, 81, 108 and 135 are also drawn; the centres of the arcs are D, G and E as
before. The oval supposed to represent the arena has an axis ratio of

$$p = \frac{\overline{OR}}{\overline{ON}} = \frac{3}{5};$$

simpler ratio values can be derived by choosing the right numbers and adding them
to or subtracting them from the three radius values in Fig. 8.9. In [15] it is suggested
that addition of 6 Roman feet yielded the oval used to trace the first step off the

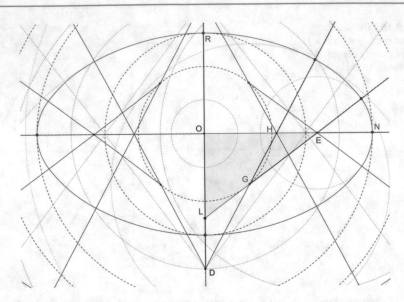

Fig. 8.11 The eight-centre oval supposed to describe the vanished arena

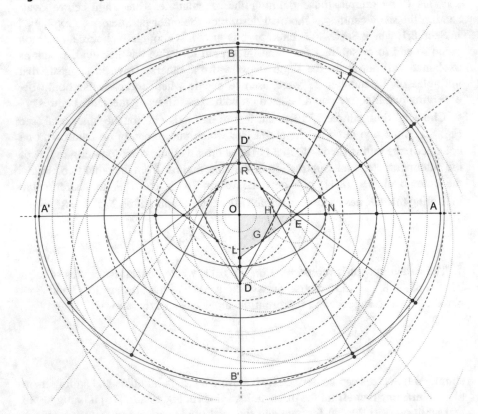

Fig. 8.12 Four eight-centre concentric ovals describing: the outer perimeter (in *blue*), the arena (in *green*), one of the ambulatory perimeters (in *red*) and the first step outside the amphitheatre (in *orange*)

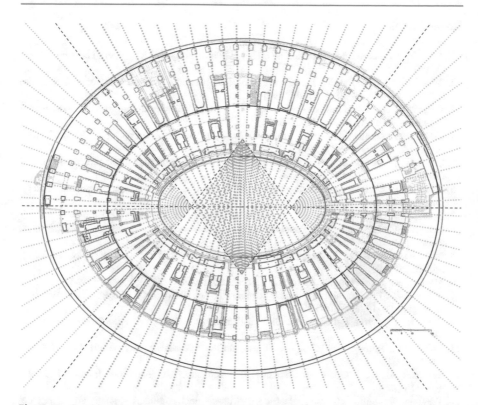

Fig. 8.13 Combination of the four-centre oval radial construction and the eight-centre concentric ovals superimposed onto the ground floor plan resulting from the 1999 survey taken from [6] *(copyright of the* Department of Representation and Survey *of the university "La Sapienza" of Rome, with kind permission)*

outer perimeter of the amphitheatre, while subtraction of 102 Roman feet yielded one of the curves describing one of the ambulatories. These two more closed curves have axis ratios respectively of

$$\frac{270}{324} = \frac{5}{6} \quad \text{and} \quad \frac{162}{216} = \frac{3}{4}.$$

The four concentric ovals are represented together in Fig. 8.12.

Superimposing Figs. 8.6 and 8.12 onto the results of the survey displayed in [6] we can get an idea of how the hypotheses on the plan of the Colosseum fit the recorded data. The result is displayed in Fig. 8.13. Note that the suggested oval for the vanished arena is not supposed to coincide with the still visible underground chamber—the hypogeum.

References

1. AA.VV: Il Colosseo. Studi e ricerche (Disegnare idee immagini X(18-19)). Gangemi, Roma (1999)
2. Benedetti, S.: Oltre l'antico e il gotico, Il profilo della cupola vaticana di Antonio da Sangallo il Giovane. Palladio. **14**, 57–166 (1995)
3. Breymann, G.A.: Trattato di costruzioni civili con cenni speciali intorno alle costruzioni grandiose. Vallardi, Milano (1926–1931)
4. Casale, A.: Alcune ipotesi sul progetto e sulle geometrie del Colosseo. Disegnare Idee Immagini. **18–19**, 81–87 (1999)
5. Croci, G.: Il comportamento strutturale del Colosseo. Disegnare Idee Immagini. **18–19**, 15–22 (1999)
6. Docci, M.: La forma del Colosseo: dieci anni di ricerche. Il dialogo con i gromatici romani. Disegnare Idee Immagini. **18–19**, 23–32 (1999)
7. Dotto, E.: Il Disegno Degli Ovali Armonici. Le Nove Muse, Catania (2002)
8. Golvin, J.-C.: L'Amphithéâtre romain: essai sur la théorisation de sa forme et de ses fonctions. Boccard, Paris (1988)
9. Mazzotti, A.A.: A Euclidean approach to eggs and polycentric curves. Nexus Netw. J. **16**(2), 345–387 (2014)
10. Migliari, R.: Ellissi e ovali. Epilogo di un conflitto. Palladio. **16**(8), 93–102 (1995)
11. Migliari, R.: Principi teorici e prime acquisizioni nel rilievo del Colosseo. Disegnare Idee Immagini. **18–19**, 33–50 (1999)
12. Ragazzo, F.: Curve Policentriche. Sistemi di raccordo tra archi e rette. Prospettive, Roma (2011)
13. Rosin, P.L., Pitteway, M.L.V.: The ellipse and the five-centred arch. Math. Gaz. **85**(502), 13–24 (2001)
14. Sciacchitano, E.: Il Colosseo. Geometria dell'impianto. Disegnare Idee Immagini. **18–19**, 107–116 (1999)
15. Trevisan, C.: Sullo schema geometrico costruttivo degli anfiteatri romani: gli esempi del Colosseo e dell'arena di Verona. Disegnare Idee Immagini. **18–19**, 117–132 (2000)
16. Wilson Jones, M.: Designing Amphitheatres, vol. 100, pp. 391–442. Mitteilungen des Deutschen Archäologischen Instituts, Römische Abteilung (1993)
17. Zerlenga, O.: La "forma ovata" in architettura. Rappresentazione geometrica. Cuen, Napoli (1997)

Appendix

A suggestion for the numbering of constructions of four-centre ovals with given axes according to the parameters known, in addition to that of Chap. 3. Non-independent groups have been left out.

Table A.1 Numbering oval constructions according to the given parameters, cases 21–52

Construction no.	Parameters			Construction no.	Parameters		
21	a	b	r_1	37	a	m	β
22	a	b	r_2	38	a	m	p
23	a	b	β	39	a	r_1	r_2
24	a	k	r_2	40	a	r_1	β
25	a	k	β	41	a	r_1	p
26	a	k	p	42	a	r_2	β
27	a	h	r_1	43	a	r_2	p
28	a	h	r_2	44	a	β	p
29	a	h	β	45	b	k	r_1
30	a	h	p	46	b	k	r_2
31	a	j	r_1	47	b	k	β
32	a	j	r_2	48	b	k	p
33	a	j	β	49	b	h	r_1
34	a	j	p	50	b	h	r_2
35	a	m	r_1	51	b	h	β
36	a	m	r_2	52	b	h	p

© Springer International Publishing AG 2017
A.A. Mazzotti, *All Sides to an Oval*, DOI 10.1007/978-3-319-39375-9

Table A.2 Numbering oval constructions according to the given parameters, cases 53–84

Construction no.	Parameters			Construction no.	Parameters		
53	b	j	r_1	69	k	h	p
54	b	j	β	70	k	j	r_1
55	b	j	p	71	k	j	r_2
56	b	m	r_1	72	k	j	p
57	b	m	r_2	73	k	m	r_1
58	b	m	β	74	k	m	r_2
59	b	m	p	75	k	m	β
60	b	r_1	r_2	76	k	m	p
61	b	r_1	β	77	k	r_1	r_2
62	b	r_1	p	78	k	r_1	r_2
63	b	r_2	β	79	k	r_1	p
64	b	r_2	p	80	k	r_2	β
65	b	β	p	81	k	r_2	p
66	k	h	r_1	82	k	β	p
67	k	h	r_2	83	h	j	r_1
68	k	h	β	84	h	j	r_2

Table A.3 Numbering oval constructions according to the given parameters, cases 85–116

Construction no.	Parameters			Construction no.	Parameters		
85	h	j	β	101	j	r_1	r_2
86	h	j	p	102	j	r_1	β
87	h	m	r_1	103	j	r_1	p
88	h	m	r_2	104	j	r_2	β
89	h	m	β	105	j	r_2	p
90	h	m	p	106	j	β	p
91	h	r_1	r_2	107	m	r_1	r_2
92	h	r_1	β	108	m	r_1	β
93	h	r_1	p	109	m	r_1	p
94	h	r_2	β	110	m	r_2	β
95	h	r_2	p	111	m	r_2	p
96	h	β	p	112	m	β	p
97	j	m	r_1	113	r_1	r_2	β
98	j	m	r_2	114	r_1	r_2	p
99	j	m	β	115	r_1	β	p
100	j	m	p	116	r_2	β	p